AF616384

TELEPEN
90 0419077 7

TOPICS IN ENZYME AND
FERMENTATION BIOTECHNOLOGY
Vol. 4

TOPICS IN ENZYME AND FERMENTATION BIOTECHNOLOGY
Vol. 4

Editor:
ALAN WISEMAN, Ph.D., F.R.I.C., M.I.Biol.
Department of Biochemistry
University of Surrey, Guildford

ELLIS HORWOOD LIMITED
Publishers Chichester

Halsted Press: a division of
JOHN WILEY & SONS
New York - Chichester - Brisbane - Toronto

First published in 1980 by

ELLIS HORWOOD LIMITED
Market Cross House, Cooper Street, Chichester, West Sussex, PO19 1EB, England

The publisher's colophon is reproduced from James Gillison's drawing of the ancient Market Cross, Chichester.

Distributors:

Australia, New Zealand, South-east Asia:
Jacaranda-Wiley Ltd., Jacaranda Press,
JOHN WILEY & SONS INC.,
G.P.O. Box 859, Brisbane, Queensland 40001, Australia.

Canada:
JOHN WILEY & SONS CANADA LIMITED
22 Worcester Road, Rexdale, Ontario, Canada.

Europe, Africa:
JOHN WILEY & SONS LIMITED
Baffins Lane, Chichester, West Sussex, England.

North and South America and the rest of the world:
Halsted Press, a division of
JOHN WILEY & SONS
605 Third Avenue, New York, N.Y. 10016, U.S.A.

British Library Cataloguing in Publication Data
Topics in enzyme and fermentation biotechnology: Vol. 4
1. Enzymes – Industrial applications
I. Wiseman, Alan
661'.8 TP248.E5 79–41466

ISBN 0–85312–150–8 (Ellis Horwood Ltd., Publishers)
ISBN 0–470–26922–7 (Halsted Press)
ISSN 0140–0835

Typeset in Press Roman by Ellis Horwood Ltd.
Printed in Great Britain by W. & J. Mackay Ltd., Chatham

Table of Contents

Chapter 1

Introduction to Topics in Enzyme and Fermentation Biotechnology 4

Dr Alan Wiseman, Biochemistry Division, Department of Biochemistry, University of Surrey, Guildford

1.1 GENERAL INTRODUCTION

This volume is the fourth in a series which was given its initial impetus by the *Handbook of Enzyme Biotechnology,* edited by Alan Wiseman and published by Ellis Horwood in 1975 (Japanese translation edition, 1977). Part 1 of that book covered the principles of enzyme production and utilisation, while Part 2 was a collection of data for use in industrial and other applications with enzymes. Many specialist topics were mentioned in this source book, although only a few were subjected to detailed analysis.

Following the *Handbook, Topics 1* (1977) reviewed enzyme synthesis in continuous culture, foam separation, aeration of culture fluids, enzymic modifications of antibiotics, patents, glucose isomerase and cytochromes P–450 *Topics 2* (1978) reviewed enzymes immobilised on inorganic supports, enzyme electrodes and enzyme based sensors, antibiotic-inactivating enzymes, biological treatment of aqueous wastes and stabilisation of enzymes. *Topics 3* (1979) reviewed the uses of oxyanions in enzyme equilibrium displacement, developments in microbial extracellular enzymes, rennets and cheese, scale up of fermentation processes and new and modified invertases and their applications.

An interdisiplinary approach is favoured in these books, as in the short post-experience courses in this field run each July at the University of Surrey 1971–1977. Many of the contributors to the series to date have also participated in these courses. The editor would be interested to hear from these participants, and others, who would perhaps like to contribute a major review to the series.

1.2 INTRODUCTION TO TOPICS IN VOLUME 4

Considerable potential exists for the exploitation of highly purified enzymes in treatment of human (and animal) diseases. A brief introduction is given to this field in Chapter 2, followed by a detailed account of the use of one class of enzymes – the proteases, by Dr Christie (Chapter 3). Some quite expensive enzymes may nevertheless be used in the future for sophisticated attempts to deliver enzymes of high specificity to appropriate intracellular sites in the body without eliciting immunological responses. Similar possibilities exist for toxic enzymes such as diphtheria toxin (see Chapter 2). Several patents have been taken out on such therapeutic enzymes, many concerned with stabilization – e.g. for an L-asparaginase entrapped in human fibrin by glutaraldehyde cross links (Y. Inada and S. Miyake for Green Cross Corp., Japan, Kokai, 77, 117, 489 dated 1/10/77): also stabilisation of urokinase with sodium dextran sulphate (H. Takayanagi, Z. Kitada, T. Sawada, A. Suzuki for Kanebo Ltd., Japan, Kokai 77, 145, 591 dated 3/10/77).

Much more work is needed on the *in vivo* aspects of enzyme therapy, including the entry of enzymes into cells by receptor-mediated endocytosis (Goldstein, Anderson and Brown, 1979).

Chapter 4, by Drs Knapp and Howell, presents an in depth review of solid substrate fermentation and its relationship to biotechnology. Aspects of fermentation are dealt with too by Drs Vincent and Priestley (Chapter 5), with much useful background information also on control of processes.

Chapter 6, by Dr Cheetham, reviews the most important and topical area of the uses of whole cells, rather than isolated enzymes, for biochemical conversions. There are advantages and disadvantages here. Whole cells are easy to obtain, but specificity, stability and freedom from contamination may be lost. Where cheapness is the main consideration the use of whole cells should dominate in certain applications; at least for a while.

REFERENCES

Goldstein, J. L., Anderson, R. G. W. and Brown, M. S. (1979) *Nature* **279**, 679–685.

Chapter 2

Enzymes in Therapy — Theory and Practice

Dr Alan Wiseman, Biochemistry Division, Department of Biochemistry, University of Surrey, Guildford

---

The high specificity and efficiency of many enzymes would suggest a worthwhile role in treatment of certain diseases. The problems lie in the relative instability of some enzymes, and in the immunological responsiveness of the body to foreign proteins especially on repeated dosage, as is usually needed. This immunological response is especially dangerous to the patient if the enzyme is allowed to enter the bloodstream. A useful alternative is the extracorporeal shunt whereby blood is pumped through an enzyme-containing device located outside the body. Even in this latter case however, responses of blood cells to this immobilised enzyme and its support may lead to the rapid inactivation of the enzyme. Oral and topical (to the skin) administration of enzymes may have fewer drawbacks therefore (see Chapter 3).

Nevertheless, enzymes will be used to restore the altered physiology and metabolism of the diseased tissue. The theory and practice of this endeavour will be discussed below.

2.1 RATIONALE FOR ENZYME THERAPY (see Fig. 2.1)

The simplest manifest need for an enzyme application is clearly in the replacement of a missing enzyme, associated usually with a congenital (inherited) disease. Very many of such deseases are known where one enzyme is missing from the appropriate cells in a particular tissue. The most intensively studied cases are the lysosome-located enzymes of the liver. This sub-cellular organelle, the lysosome, contains a whole range of enzymes needed to degrade macromolecules, and the absence of one enzyme type may eventually prove fatal. The lysosome enzymes are however usually glycoproteins, containing a carbohydrate moiety therefore. Efficient incorporation of exogeneously administered enyme, even in the special lipid packages (liposomes: see Gregoriadis, 1977) used, may not be possible with microbial enzymes. Human enzymes are often used when available. Placenta is a valuable source of appropriate human enzymes at present. Human cells culture may also be developed, but no doubt in future, genetic engineering will lead to the production of the appropriate human enzyme by fermentation, using microorganisms.

Another obvious application of an enzyme of appropriate specificity is in the removal, usually from the bloodstream, of an undesirable substance that may be causing symptoms. Thus urea can be removed with urease, uric acid with uricase or urate oxidase, and oxalic acid with oxalate decarboxylase. In addition the possibility exists that the enzyme could be released slowly, or after a latent period, by for example the degradation of an attached polymer.

Given sufficient knowledge of the biochemical control of the body it should prove possible to elicit more sophisticated responses such as switch on or off effects. Another possibility is an effect on cell surfaces (and cellular contact) for example to get tumour cells to be more easily recognised as foreign by the immunological defences of the body.

The most exciting development in recent years has been in the treatment of cancer. Certain tumour cells may require an unusual nutrient from the bloodstream, or a usual nutrient but at higher

RATIONALE FOR ENZYME THERAPY

a) To *replace* a missing enzyme — congenital (inherited) disease.

b) To *remove* an undesirable substance that is causing symptoms, e.g. urea.

c) To *deplete* a nutrient required specifically by a tumour cell (cancer).

therefore

To *restore* the altered physiology and metabolism of the diseased tissue.

ENZYMES OFFER; SPECIFICITY and EFFICIENCY.

PROBLEMS: DELIVERY: STABILITY: IMMUNOLOGICAL RESPONSE.

Fig. 2.1

concentration than normal cells. The possibility exists therefore of preventing the continued growth of the tumour by restricting its special nutrient supply, by destroying this with an administered enzyme. In this area, much published work has appeared on the use of microbial asparaginases (and glutaminases) in the treatment of some forms of leukaemia. Very considerable effort has gone into the development of suitable enzymes and in understanding their value and their disadvantages. Many problems exist still with these treatments, but considerable success has been achieved especially in combination with the use of the appropriate cytotoxic drug.

Consideration will now be given to the necessary features of the enzymes to be employed, followed in later sections by the details of particular enzyme applications in therapy.

2.2 ENZYMIC REQUIREMENTS FOR THERAPY (see Fig. 2.2)

The most obvious requirement must be met. Sufficient enzyme with appropriate activity must be available for the intended course of treatment. In very many cases there is insufficient enzyme for the purpose, especially in correctly organised clinical trials. Despite therefore the success of the enzyme in any appropriate animal model (if available) of the disease to be treated, little progress can be made until production of the required enzyme is initiated on the scale really necessary. The enzyme must nevertheless be produced with sufficiently high specific activity for the intended purpose, and in

REQUIREMENTS FOR ENZYME THERAPY

Enzyme of appropriate substrate-specificity, purity/source, cost/quantity, stability, delivery-form worth developing.

Stabilisation by cross-linking, immobilisation to insoluble or soluble supports, encapsulation (to prevent denaturation, degradation, inactivation/inhibition).

Delivery-form liposomes, microcapsules, red blood cells (also shunts). Lower immunological response. Target-directed affinity.

Fig. 2.2

addition should have high enough purity not to elicit unwanted side reactions not due to the enzyme itself. These are usually due to other proteins present, including microbial endotoxins. Organisms pathogenic to man are therefore a most unlikely source of useful enzyme, unless a most extensive purification procedures can be undertaken. These will of course make the enzyme, and the treatment, even more expensive – perhaps prohibitively so.

Prior studies on experimental animals may have established the likely usefulness of the enzyme, based upon the balance between efficacy and toxic side effects. A high activity and stability for the enzyme, at physiological pH, is essential. This must be equally so in the whole blood of the animal, and of man, if the enzyme is to be injected in free form. The ability of the enzyme to reach its normal, or correct, site of activity must be ascertained, and its usefulness demonstrated despite any immunological responses, especially on repeated dosing.

From an enzyme kinetics viewpoint an appropriate enzyme will require certain features, in addition to stability and correct pH optimum. These include a reasonably high affinity (low K_m e.g. 10^{-5} M for asparaginases; Howard and Carpenter, 1972) for the substrate, viewed in relation to the substrate concentration in blood or tissue. A high rate of conversion per molecule of enzyme (turnover number) at that substrate concentration is desirable, to save on quantity of enzyme needed for useful therapeutic effect. Other useful features should include the absence of inhibition of the enzyme by its products or other substances present, and a reaction that effectively goes to completion under the conditions of use. Many enzymes require co-enzymes and co-factors such as adenosine triphosphate (ATP) and specific metal ions. These are usually less useful for therapy purposes. Possibilities of regeneration of co-enzymes exist, especially in a multi-enzyme immobilised enzyme package which may possibly be used, especially in the form of extracorporeal shunts for the blood. Another constraint where the enzyme must be effective in the bloodstream is in the pharmacokinetics of retention of the enzyme there. Only a *slow* clearance of the enzyme from circulation may be essential for efficacious therapy. Possibilities exist here of encapsulation in permeable microcapsules. Alternatively, chemical modification of the soluble enzyme can be achieved so as to increase the plasma half-life of the enzyme. This may occur perhaps by protection against degradation by proteases present, or may be related to the ease of removal of the enzyme by phagocytosis. A correlation has been suggested with this and the isoelectric point in the case of asparaginase (Rutter and Wade, 1971).

2.3 SOME THERAPEUTIC POSSIBILITIES (see Fig. 2.3)

Recent interest in the use of enzymes in therapy has been concerned with the treatment of patients with some forms of leukaemia. For example, L-asparaginases have been isolated and purified from micro-organisms *(E. coli* or *Erwinia carotovora)* and used in this treatment (for general review see Cooney and Rosenbluth, 1975). The basis of the treatment is the degradation of the plasma amino acid, asparagine, that is essential for the survival of the tumour cells. High rates of remission have been reported for this treatment, especially when combined with the use of cytotoxic drugs (Tang, 1970). Remission may be for several months in some patients, although hypersensitivity reactions are more common during second and subsequent courses of therapy. Immobilised forms of the enzyme (including microcapsules) have been investigated, but the specific activity of the enzyme preparations available has been lower than optimal. Fully controlled clinical trials will be necessary to find the best available forms of treatment with these enzymes. L-asparaginase has been used also as an immunosupressive agent in the treatment of organ allograft rejection and connective tissue disorders (Hersh, 1971).

Several fibrinolytic enzymes are available now. Streptokinase, from haemolytic streptococci, and urokinase, from human urine, given intravenously, initiate the activation of plasminogen to plasmin

SOME THERAPEUTIC POSSIBILITIES

Specific substrate-dependent tumours, e.g. asparaginase/glutaminase. LEUKAEMIA

Hereditary enzyme-deficiency diseases, e.g. large number of lysosomal storage diseases.

Organ failure and metabolic disorders, e.g. urease; uricase; oxalate-decarboxylase; Cytochrome p-450.

Substitutes for blood cells.

Blood clot removal.

Inflammatory conditions.

NOTE: Important role of experimental therapy in animals, using disease models.

Fig. 2.3

and thus help to dissolve blood clots in cases of thrombosis (review: Tang, 1970). Urokinase has many advantages over streptokinase as a thrombolytic agent, including absence of streptoccal antibody problems. Another possibility is the removal of fibrinogen by the action of enzymes derived from snake venoms(Arvin; Reptilase). The clots formed by the action of these enzymes are more easily dissolved by plasmin than the thrombin-induced clot (Ewart, *et al.*, 1970). Another suggestion in this area, is in the use of brinase, isolated from *Aspergillus oryzae*.

Many other proteolytic enzymes have proved useful in some cases of inflammatory conditions. These include bromelain from pineapple stem, toxin-free collagenase from *Clostridium histolyticum* (and *Cl. welchii*), hyaluronidase from snake venom or pig testes, papain from the papaya melon, and trypsin/chymotrypsin (Chymoral) from beef pancreas. Some of these enzymes are administered orally (Margetts, *et al.*, 1972; see Chapter 3).

Other enzymes of interest include lysozyme from egg white for eye therapy, penicillinase from *Bacillus cereus* to terminate penicillin allergy, and deoxyribonuclease from beef pancreas, to help dissolve blood clots.

Recent developments include the attempted treatment of congenital diseases by enzyme replacement therapy. Target organ-directed delivery of such enzymes may be possible by use of specific-surface liposomes (Juliano and Stamp, 1976). Trials have included the attempted treatment of Tay Sachs disease (G_{M2} gangliosidoses type 2) with urinary hexosaminidase A; type II glycogenosis with placental α-glucosidase; Fabry's disease with placental α-galactosidase A (ceramidetrihexosidase); and Gaucher's disease type 1 with placental β-glucosidase.

An important review has been published on enzymes as drugs (pharmacological aspects) by Holcenberg and Roberts (1977).

Recent interest has been noted in several newer enzymic possibilities. One area is in the treatment of cancer through enzymic depletion of folic acid. Carboxypeptidase G_1 is a folate-hydrolysing enzyme (cleaves C-terminal glutamyl residues) isolated from *Pseudomonas stutzeri* that had been shown earlier to inhibit the growth of human lymphoblastoid cells *in vitro* (Bertino, *et al.*, 1972). The same enzyme hydrolyses the drug, methotrexate, the folate inhibitor often used in therapy. Excess (toxic) methotrexate can be removed with carboxypeptidase a day or two after administration. Another carboxypeptidase, from *Acinetobacter spp.* may be particularly useful in this role (Albrecht, *et al.*, 1976). The same workers have isolated a folic acid deaminase from *Pseudomonas spp.* (Albrecht, *et al.*, 1974).

Other enzymes of interest include a L-phenylalanine ammonia lyase from *Rhodoturola glutinis* (Abell, *et al.*, 1972). This enzyme depletes phenylalanine and tyrosine by deamination to the corresponding trans-cinnamic acid. Removal of these amino acids may slow tumour growth in some favourable cases.

An interesting enzyme application is in the modification of tumour cells *in vitro* using neuraminidase from *Vibrio cholerae*. The enzyme cleaves the terminal N-acetyl-neuraminic acid (sialic acid) residues from the glycoproteins located on the surface of the cell. These modified tumour cells elicit a greater immunological response to the tumour by the host when injected back into the patient (Holland and Bekesi, 1976). The enzyme has proved effective in immobilised form (Bazarian and Wingard, 1978).

2.4 USE OF MODIFIED FORMS OF ENZYMES

Several types of modification are possible, including chemical modification of functional groups, cross-linking, attachment of soluble polymers, attachment of insoluble polymers (immobilisation) – or a combination of two or more of these methods. The aim is stabilisation, ease of utilisation, modification of enzymic properties such as pH optimum, modification of enzyme kinetic properties such as K_m, and prolongation or control of tissue incorporation or plasma half-life. The other possibility is that the immunological response to the enzyme may be decreased.

Unique properties have been claimed for enzymes joined to polyethylene glycol (Davis, *et al.*, 1978). Such adjuncts have been prepared with four tetrameric enzymes, bovine liver catalase, bovine liver arginase, hog liver uricase and *Candida utilis* uricase. These enzyme adjuncts were non-immunogenic in the rabbit after intramuscular injection. Covalent attachment of polyethylene glycol to the amino groups of catalase has been noted as a method of decreasing the immunological response of this enzyme (Abuchawski, *et al.*, 1974).

2.4.1 Chemical modification of soluble enzymes

Asparaginases have been modified in a variety of ways to increase their plasma half-life. Firstly, studies on the partially deaminated enzyme was reported by Wagner, *et al.* (1969). The enzyme was deaminated with nitric acid, at the N-terminal leucine and one ϵ-amino of lysine, without loss of enzymic activity. Holcenberg, *et al.* (1975) have reported that succinylation of the amino groups of the glutaminase-asparaginase complex from *Acinetobacter spp.*, which converts the amino group to a carboxylic acid functional group, greatly increases the plasma half-life in experimental animals. In addition, they demonstrated that the susceptibility of the enzyme to tryptic digestion was greatly decreased *in vitro*. These enzymes, however, are likely to be destabilised by removal of most of the amino groups, suggesting the possibility of stabilisation by salt-links as has been claimed by Woodward and Wiseman (1978) for invertase.

Other derivatives of L-asparaginase have included reactive p-benzoquinone adducts which were coupled to human erythrocytes. Other reactive derivatives included m-maleoamidobenzoate and N-maleoyl amino acid adducts (Malterella and Richardson, 1979).

Cross-linking of enzymes usually increases their thermal stability, presumably by decreasing the extent of conformation change due to random thermal motion (for review on enzyme stabilisation see Wiseman, 1978). Some loss of specific activity may result, however, for example, in the use of dimethyl suberimidate to cross-link the lysine residues of asparaginase from *E. coli* (Handschumacher and Gaumond, 1972). Nitroasparaginase made by the cross-linking of tyrosine residues with tetranitromethane displayed altered substrate specificity (Liu and Handschumacher, 1972).

Glutaraldehyde is often used as a cross-linking reagent. For example, uricase and albumin have been treated with glutaraldehyde and the complex was shown to have a long half-life in dogs, with good stability of the uricase activity Poznonsky, 1977). This and many other examples are quoted in the books by Chang (1977) on biomedical uses of enzymes.

Conjugation of enzymes to carbohydrates such as soluble dextran have often improved their stability (see Wiseman, 1978).

2.4.2 Immobilised enzymes in therapy

Various possibilities exist for extracorporeal shunts (blood), implants and prostheses. Immobilisation of the enzyme is a useful method of containing and possibly stabilising the enzyme. The solid support, for insoluble immobilised enzymes, can range from living tissue to glass beads, fibres or the walls of tubes or dialysis membranes. The enzyme may be chemically attached, or entrapped in small lipid globules (liposomes) or for example gelatin or collagen microcapsules or mini-cells. Red blood cells themselves may be readily emptied and refilled with enzyme solutions through adjustment of salts concentration (Ihler, *et al.,* 1973). These are rather impermeable however, and would discharge their enzyme inside the reticuloendothelial system. Some of these encapsulated forms are in ultrathin membranes and are readily permeable to substrate in the bloodstream or site of retention. Others such as liposomes need to discharge their contained enzyme, in this case inside the cellular lysosome following entry into the cell.

It should be possible to direct such liposomes and capsules to specific sites in the body, or even to tumours, by labelling their surface with an appropriate antibody, produced in experimental animals. Tumours may even be attacked with toxins this way, such as the enzyme, diphtheria toxin. This can be produced from *Corynebacterium diphtheriae* and is an enzyme that uses NAD^+ and catalyses the ADP-ribosylation (with splitting of NAD^+ and liberation of nicotinamide) of an enzyme required in protein biosynthesis (review, Collier, 1975). This enzyme in protein biosynthesis is usually called elongation factor 2 in animals, and G factor or (GTP-dependent) translocase in bacteria. Controlled and specific liberation of this dangerous toxin at tumour cell surfaces would be a possibility.

2.4.2.1 PREPARATION AND USE OF MICROCAPSULES

T.M.S. Chang (see books, 1977) has pioneered this most useful approach whereby microdroplets of aqueous enzyme solution are entrapped in ultrathin and permeable polymer membrane (e.g. nylon, gelatin and others including *in situ* polymerisation techniques). Several interesting variants are possible in these methods of preparation of therapeutically useful microcapsules and in addition it is possible to include within the microcapsule other materials to facilitate the enzymic reaction (see on). In addition, the enzyme doesn't escape easily into the circulation, thus reducing immunological responses and damage to the blood cells.

Urease has been used (in dogs), with an absorbent for the ammonia produced, in removing urea from the bloodstream, so as to remove this toxic material and facilitate also the enzymic reaction itself (Chang, 1966). Fast enzymic conversion is possible, with large surface area of microcapsules, and stabilisation of the microencapsulated enzyme may be improved further, in addition to proteolysis protection (Chang, 1971b) by immobilisation and cross-linking. This can be achieved also by the simple expedient of increasing the protein concentration, for example, with 10% haemoglobin (Chang, 1971a). Experiments in mice have demonstrated the use of encapsulated catalase (Chang and Poznonsky, 1968) and more recently of encapsulated asparaginase in possible treatment of leukaemia (Chang, 1971b) and in achieving regression of an implanted lymphosarcoma (Chong and Chang, 1974).

More recently the use of nylon microsphere encapsulated deoxyribonuclease has been demonstrated, in extracorporeal shunt in the dog. This was used as a possible treatment for the autoimmune disease, systemic lupus erythomatosus, where circulating DNA may be a problem (Terman, 1976). Microcapsules may be used by intravenous injection and in some cases by intraperitoneal injection, but this may cause other problems. Immunological responses to the capsules may cause their clumping together with a lowering of their efficiency. The efficiency may be reduced anyway by the usual deleterious effects of diffusion on enzyme kinetic parameters such as K_m and V_{max}.

2.4.2.2 PREPARATION AND USE OF LIPOSOMES

Liposomes are lipid capsules containing lipid bilayers in alternation with aqueous layers, in which enzyme solution can be incorporated (Gregoriadis, *et al.*, 1972, 1977). It has proved possible to incorporate enzymes in the lipid bilayers too, where appropriate.

Liposomes are made by use of various mixtures of lipids such as lecithin (egg phosphatidyl cholines) and cholesterol. A positive surface charge can be achieved by incorporation of octadecylamine (sterylamine), and/or a negative surface charge by use of dicetyl phosphate. Surface charge determines the plasma half-life of these liposomes. These lipids are dissolved in an organic solvent and rotary evaporated to produce a thin film on the vessel wall. An aqueous buffered solution of the enzyme is added and the liposomes form with gentle shaking. These are relatively large (about one millimetre in diameter) but can be made much smaller by ultrasonication treatment. The rate of uptake, and the site of uptake, of liposomes may be determined partially by the size (Juliano and Stamp, 1975). Removal of entrapped soluble enzyme is usually accomplished by gel filtration, although centrifugation may be used to fractionate the liposomes (Heath, *et al.*, 1976), or achieve a sterile pyrogen-free product for therapy (Tyrrell and Ryman, 1977). Other methods of preparation have been reported including injection through fine needles (Batzri and Korn, 1975).

Papadopoulos (1978) has written a useful review of the present and potential uses of enzymes in therapy, and has included an account of liposome therapy. In addition the stability of various types of liposome was investigated using yeast and liver alcohol dehydrogenases (Papadopoulos and Wiseman, unpublished). The enzymic activity was not accessible to alcohol in this investigation but could be made so by treatment with the detergent, Triton X-100. Apparent thermal stability of the entrapped enzyme was much improved.

Very many different enzymes have now been entrapped in liposomes of various compositions, but one disappointment is that immunological responses still occur, and these may be greater than for free enzyme (due to an adjuvant effect), especially upon repeated dosage (Allison and Gregoriadis, 1974). The liposomes are eventually taken up, mainly into the liver and spleen by endocytosis mechanisms, and the enzyme is liberated, usually in the lysosome sac containing degradative enzymes. The missing enzyme in liver lysosomes in, for example, a glycogen storage disease might be replaced this way (Tyrrell and Ryman, 1977). Liposomes may be useful too for oral administration of enzymes and in various ways for administration of drugs.

A useful brief review has been written by Gregoriadis and Dean (1979) on enzyme therapy in genetic diseases. Several areas of advance were indicated, especially as many enzymes possess recognition markers that facilitate their rapid and specific uptake by cells. For example, many lysosomal hydrolases have a phosphomannose group that mediates the uptake of the enzyme by fibroblasts. Direction to particular tissues may be achieved by modification or control of such carbohydrate sites in glycoprotein enzymes. Nevertheless, good animal models for human genetic diseases are difficult to find, and immunological responses may not be predicted anyway, on animals, when using human enzymes.

2.4.2.3 PREPARATION AND USE OF ENZYME LOADED ERYTHROCYTES

As mentioned above, the red blood cell can be emptied and refilled by reversible osmotic shock (Ihler, *et al.*, 1973). The cells are washed several times with buffer to remove plasma, and then the enzyme solution (in dilute buffer) is added. Within less than a minute it is necessary to stop the erythrocytes from bursting (haemolysis) by the addition of sodium chloride solution to a final concentration of 0.9%. Almost all the enzyme has entered the red cells and cannot be washed out, except after bursting the cells in distilled water, by osmotic forces.

The enzyme-filled erythrocytes may be active in the bloodstream, when reintroduced, to remove substrates of the enzymes. Also the red cell can be taken up by other blood cells (phagocytes) where an appropriate enzyme might be of therapeutic value. The loaded erythrocytes may be quickly taken up by the reticuloendothelial system and the enzyme released there. Such a system has been tried in β-glucuronidase-deficient mice using this enzyme inside erythrocytes

(Thorpe, *et al.*, 1975). In Gaucher's disease, the accumulation of β-glucocerebroside in cells of the reticuloendothelial system (including spleen) may perhaps be treated with the β-glucocerebrosidase (Ihler, *et al.*, 1973). These authors have studied uricase too *in vitro* (Ihler, *et al.*, 1973).

Another example is in the entrapment of enzymes and drugs in resealed erythrocyte 'ghosts' prepared from the blood of rat, mouse or man. Invertase was used, and the fate of the intravenously injected erythrocytes was followed by labelling with the γ-emitter 99m T^c (Tyrrell and Ryman, 1977).

2.4.2.4 SOME USES OF IMMOBILISED ENZYMES

There have been many attempts to decrease the immunogenicity of asparaginase, for example by chemical modification (see 2.4.1). Immobilisation may have this effect also, by preventing the uptake of the enzyme by lymphocytes. Asparagine may nevertheless be removed from the bloodstream with appropriately immobilised asparaginase, in some instances by extracorporeal shunt devices.

For example, this enzyme has been immobilised to nylon tubing with glutaraldehyde and used to treat leukaemic dogs (Allison, *et al.*, 1972), and on polymethylmethacrylate (Horvath, *et al.*, 1973) attached to nylon tubing. The enzyme was thereby securely fixed and protected from proteolytic enzymes and antibodies. Diffusion limitation of substrate access was the usual problem encountered, however. Another problem is the accumulation of fibrin deposits, and plasma proteins especially in the bloodstream. This occurred with the Dacron vascular prosthesis prepared by Cooney, *et al.* (1975). The enzyme was attached by silanisation of the surface and reduction and diazotisation. When such preparations were injected into mice bearing ascitic tumours, an immunological response occurred due to some leakage of the enzyme from the polymer because of breakage of covalent bonding. Similar leakage of asparaginase was noted by Sampson, *et al.*, (1974) off glass plates following immobilisation of the enzyme by silanization and glutaraldehyde coupling of the enzyme. Other studies (O'Driscoll, *et al.*, (1975) used gel-entrapped *E. coli* asparaginase, using polyhydroxyethyl methacrylate. Salmona, *et al.* (1974) used fibre-entrapped asparaginase (cellulose triacetate fibre) in mice peritoneal cavity, to study antibody formation. Immunosupressive effects of asparaginase were observed. More recently, Lange, *et al.* (1977) reported the use of an extracorporeal device with immobilised asparaginase, in treatment of melanoma.

The author's interests in drug-metabolising yeast cytochrome P-450 from yeast (Woods and Wiseman, 1979) include another medical application of immobilised enzyme related to drug-metabolism by artifical liver systems. Several such systems have been proposed to cope with organ failure, and detoxifiers to treat drug overdose or liver failure would be useful in therapy (Chang, 1977). A membrane immobilised liver microsomal fraction has been designed for this purpose (Cohen, *et al.*, 1977). Solubilised and purified rabbit microsomal enzymes were used by Brunner and Losgen (1976). These workers used lipase to solubilise enzymes including UDP-glucuronyl transferase, which is then immobilised. This was tested successfully in rabbits, but it was necessary to load the solid support with the relatively expensive co-enzyme UDP-glucuronic acid. Solid supports tested included acrylamide-acrylic acid-imide copolymers, and Sepharose derivatives. Not all carriers were considered suitable for contact with blood, as granules had a tendency to destroy erythrocytes and blood platelet cells. Membrane carriers may be better in this respect. Other useful forms of immobilised enzyme, especially for extracorporeal shunts, will include tubular bundle reactors and hollow fibre capsules. More than seventeen papers were quoted in the review by Sofer (1979) on artificial liver systems. One of the main problems is in the regeneration of expensive, and toxic, coenzyme (NADPH). Coimmobilisation of co-enzyme regenerating enzymes may be essential. Immobilised co-enzymes have been developed too (Mosbach, Larsson and Lowe, 1976).

Other reports include the development of an immobilised glutathione transferase (ligandin) by Brunner (1978) and of a hollow-fibre entrapped liver microsome system for use as an extracorporeal shunt for detoxication (Kastl, *et al,* 1978).

2.5 CONCLUDING COMMENTS

The not insignificant use of soluble enzymes in therapy will no doubt be extended further, and the new opportunities with immobilised enzymes should allow a variety of novel devices such as membrane oxygenators containing carbonic anhydrase (Brown, *et al.*, 1970), and fuel cell cardiac pacemakers containing glucose oxidase (Drake, 1968). Protection against radiation damage may be possible using superoxide dismutase (Petkau, 1979).

As with all new therapies, adequate testing for usefulness and toxicity in experimental animals is essential, especially in disease model systems if available. Results from fully controlled clinical trials should be of great inrerest with many of the new systems – see Chapter 3 by Dr Christie.

REFERENCES

Abell, C. W., Stith, W. J. and Hodgins, D. S. (1972) *Cancer Res.*, **32**, 285.

Abuchawski, A., Van Es, T., Pulczuk, N. C. and Davis, F. F. (1974) *Fed. Proc.*, **33**, 1317 (Abstract).

Albrecht, A. M., Boldizsar, E. and Hutchinson, D. J. (1976) *Fed. Proc.*, **35**, 787 (Abstract).

Albrecht, A. M., Fenton, J. J., Zanati, E. and Hutchinson, D. J. (1974) *Fed. Proc.*, **33**, 715 (Abstract).

Allison, J. P., Davidson, L., Gietierrez-Hartman, A. and Kitto, G. B. (1972) *Biochem. Biophys. Res. Commun.*, **47**, 66–73.

Allison, J. P. and Gregoriadis, G. (1974) *Nature*, **252**, 252.

Bazarian, E. R. and Wingard, Jr. L. B. (1978) In: *Enzyme Engineering*, **4**, (ed. Brown, G. B., Manecke, G. and Wingard, Jr. L. B.), Plenum Press, New York and London. 211–212.

Batzri, S. and Korn, E. D. (1975) *J. Cell Biol.*, **66**, 221–224.

Bertino, J. R., O'Brien, P. and McCullough, J. R. (1972) *Science*, **172**, 161–162.

Brown, G., Selegny, E., Minh, C. T. and Thomas, D. (1970) *FEBS Lett.*, **7**, 223.

Brunner, G. (1978) In: *Enzyme Engineering*, **4**, (ed. Broun, G. B., Manecke, G. and Wingard, Jr., L. B.), Plenum Press, New York and London, 175–177.

Brunner, G. and Losgen, H. (1976) In: *Enzyme Engineering*, **3**, (ed. E. K. Pye and H. H. Weetall), Plenum Press, New York and London, 391–396.

Cohen, W., Baricos, W. H., Kastl, P. R. and Chambers, R. P. (1977) In: *Biomedical Applications of Immobilised Enzymes and Proteins*, Vol. 1, (ed. T. M. S. Chang), pp. 319–328.

Collier, R. J. (1975) *Bacteriol. Rev.*, **39**, 54–85.

Cooney, D. A. and Rosenbluth, R. J. (1975) *Adv. in Pharm & Chemother*, **12**, 185–289.

Cooney, D. A., Weetall, H. H. and Long, E. (1975) *Biochem. Pharmacol.*, **24**, 503–515.

Chang, T. M. S. (1977) *Biomedical Applications of Immobilised Enzymes and Proteins*, Vols. 1 & 2, Plenum Press, New York.

Chang, T. M. S. (1971b) *Nature*, **229**, 117–118.

Chang, T. M. S. (1966) *Canad. J. Physiol. Pharmacol.*, **44**, 115.

Chang, T. M. S. (1971a) *Biochim. Biophys. Res. Communs.*, **44**, 1531–1536.

Chang, T. M. S. and Poznonsky, M. J. (1968) *Nature*, **218**, 243–245.

Chong, E. D. S. and Chang, T. M. S. (1974) *Enzymes*, **18**, 218.

Davis, F. F., Abuchawski, A., Van Es, T., Palczuk, N. C., Chen, R., Savoca, K. and Wieder, K. (1978) In: *Enzyme Engineering,* Vol. 4, (ed. Broun, G. B., Manecke, G. and Wingard, Jr., L. B.), Plenum Press, New York and London, pp. 169–173.

Drake, R. F. (1968) U.S. Government Report, PB 177695.

Ewart, M. R., Hatton, M. W. C., Basford, J. M. and Dodgson, K. S. (1970) *Biochem. J.,* **118,** 603.

Gregoriadis, G. (1977) *Nature,* **265,** 407–411.

Gregoriadis, G. and Dean, M. F. (1979) *Nature,* **278,** 603–604.

Gregoriadis, G. and Ryman, B. E. (1972) *Europ. J. Biochem.,* **24,** 485–491. see also FEBS Lett., (1977) **14,** 95–99).

Handschumacher, R. E. and Gaumond, C. (1972) *Mol. Pharmacol.,* **8,** 59–64.

Heath, T. D., Edwards, D. C. and Ryman, B. E. (1976) *Biochem. Soc. Trans.,* **4,** 129–133.

Hersh, E. M. (1971) *Transplantation,* **12,** 368.

Holcenberg, J. S. and Roberts, J. (1977) Ann. Rev. Pharmacol. Toxicol., **17,** 97–116.

Holcenberg, J. S., Schmer, G., Teller, D. C. and Roberts, J. (1975) *J. Biol. Chem.,* **250,** 4165–4170.

Holland, J. F. and Bekesi, J. G. (1976) *Med. Clin. North America,* **60,** 539.

Horvath, C., Sardi, A. and Woods, J. S. (1973) *J. Appl. Physiol.,* **34,** 181–187.

Horward, J. B. and Carpenter, F. H. (1972) *J. Biol. Chem.,* **247,** 1020.

Ihler, G. M. (1973) *Proc. Natl. Acad. Sci.,* U.S., **70,** 2663–2666.

Ihler, G. M. (1973) *J. Clin. Invest.,* **56,** 595–602.

Ihler, G. M., Glew, R. H. and Schnure, F. W. (1973) *Proc. Natl. Acad. Sci.,* U.S., **70,** 2663–2666.

Juliano, R. L. and Stamp, D. (1975) *Biochem. Biophys. Res. Communs.,* **63,** 651–658.

Juliano, R. L. and Stamp, D. (1976) *Nature,* **261,** 235.

Kastl, P. R., Baricos, W. H., Chambers, R. P. and Cohen, W. (1978) In: *Enzyme Engineering,* **4,** (ed. Broun, G. B., Manecke, G. and Wingard, Jr., L. B.), Plenum Press, New York and London, 199–206.

Lange, A. *et al.* (1977) *Scand. J. Haematol.,* **18,** 244–248.

Liu, Y. P. and Handschumacher, R. E. (1972) *J. Biol. Chem.,* **247,** 66–69.

Malterella, N. and Richardson, T. (1979) *Biochem. Soc. Trans.,* **7,** 66–69.

Margetts, G., Barber, K., Christie, R. B., Jones, W. E. and Bowden,

W. T. (1972) *Brit. J. Clin. Pract.*, **26**, 293.

Mosbach, K., Larsson, P.-O. and Lowe, C. R. (1976) In: *Methods in Enzymology,* Vol. 44, (ed. Mosbach, K.), Academic Press, pp. 859–887.

O'Driscoll, K. F., Korus, R. A., Ohnuma, T. and Walczack, I. M. (1975) *J. Pharmacol. Exp. Ther.*, **195**, 382–388.

Papadopoulos, P. (1978) Project Dissertation, Department of Biochemistry, University of Surrey, pp. 1–145.

Petkan, A. (1979) *Photochemistry and Photobiology,* **28**, 765.

Poznonsky, M. J. (1977) In: *Biomedical Applications of Immobilised Enzymes and Proteins,* (ed. Chang, T. M. S.), Plenum Press, New York, pp. 341–354.

Rutter, D. A. and Wade, H. E. (1971) *Brit. J. Exptl. Pathol.*, **52**, 610.

Salmona, M., Saronio, C. and Garattini, S. (1974) *Insolubilised Enzymes,* Raven Press.

Sampson, D., Han, T., Hersh, L. S. and Murphy, G.P. (1974) *J. Surg. Oncol.*, **6**, 39–48.

Sofer, S. S. (1979) *Enzyme Microb. Tech.*, **1**, 3–5.

Tang, P. (1970) *Process Biochemistry,* **5**(2), 46.

Terman, G. (1976) *J. Clin. Invest.*, **57**, 1201–1212.

Thorpe, S. R., Fiddler, M. B. and Desnick, R. J. (1975) *Pediatr. Res.*, **9**, 918–923.

Tyrrell, D. A. and Ryman, B. E. (1977) *Biochem. Soc. Trans.*, **4**, 677–681.

Tyrrell, D. A. (1976) *Biochim. Biophys. Acta,* **457**, 259–302.

Wagner, O., Irion, E., Arens, A. and Bauer, K. (1969) *Biochem. Biophys. Res. Commun.*, **37**, 383–392.

Wiseman, A. (1978) In: *Topics in Enzyme & Fermentation Biotechnology,* Vol. 2, (ed. A. Wiseman), Ellis Horwood (Wiley), pp. 280–303.

Woods, L. F. J. and Wiseman, A. (1979) *Biochem. Soc. Trans.*, **7**, 124–125, see also Wiseman, A., Lim, T.K. and Woods, L. F. J. (1978) *Biochim. Biophys. Acta,* **544**, 615–623.

Woodward, J. and Wiseman, A. (1978) *Biochim. Biophys. Acta,* **527**, 8–16.

Chapter 3

The Medical Uses of Proteolytic Enzymes

Mr R B Christie, Armour Pharmaceutical Company Ltd.,
Eastbourne, Sussex

3.1 INTRODUCTION

Reference to the medical use of proteolytic enzymes stretches back for well over 100 years. Early explorers of the Caribbean islands noted that the local inhabitants used fresh pineapple juice as a digestive aid. Native remedies to improve the texture of their skin, and to remove necrotic tissue, involved the application of crude papain, a proteolytic enzyme isolated from the paw-paw fruit, *Carica papaya.* Bromelain in the form of pineapple juice was also utilised for this purpose.

Proteolytic enzymes secreted from the mouth parts of live maggots were customarily applied in early military and naval surgical medicine to clean suppurating wounds sustained as a result of battle casualties.

Pepsin, and pharmaceutical preparations of this enzyme, were described in several early pharmacopoeias including the *British Pharmacopoeia* of 1898. Methods of preparing various elixirs and elementary standards for pepsin as an aid to digestion were thus established.

A few relatively impure enzyme preparations, such as pancreatin, remain in use today, but the emphasis, as with most naturally occurring drugs, has been to separate and purify for a specific activity and clinical use.

The final logical stage in this sequence of events, as illustrated by the peptide hormones, is the synthesis of the active molecule or at least an active fragment. This has not proved practable, as yet, for proteolytic enzymes because of the large size of the molecule. For example, the molecular weight of trypsin (porcine) is 23 400 (Travis and Liener, 1965), chymotrypsin (bovine) 25 310 (Eck and Dayhoff, 1966), and pepsin (porcine) 36 000 (Cunningham, 1965).

The therapeutic use of proteolytic enzymes is normally coupled to their ability to selectively degrade proteins to simpler peptides and amino acids, although this generality may not apply in all cases, as for example their use as anti-inflammatory agents.

Many different proteolytic enzymes are involved, their precise application being a function of their substrate specificity, low local and systemic toxicity, and the possibility of compounding them into an appropriate pharmaceutical formulation for the particular clinical indication concerned.

A brief review of the specific enzymes involved and their general properties will provide an introduction to a description of their clinical application.

3.2 PROTEOLYTIC ENZYMES USED IN MEDICINE – GENERAL PROPERTIES

The general properties of proteolytic enzymes are well documented in other publications. It is only proposed to outline those which are directly relevant to the preparation, standardisation and use of these enzymes in medicine.

3.2.1 Trypsin

Trypsin is separated from an acid extract of mammalian pancreas as the inactive pre-enzyme Trypsinogen, which is further purified by ammonium sulphate fractionation. The pre-enzyme may be activated by small quantities of trypsin at pH 8, or by enterokinase. Isolation was first reported by Northrop and Kunitz (Northrop and Kunitz, 1932).

Trypsin hydrolyses the peptide linkages in proteins and partly hydrolysed proteins, with preferential attack on lysine and arginine residues. In most instances, it has only a weak action on native proteins such as haemoglobin, serum proteins, or collagen.

It is inhibited by a wide variety of inhibitors; from a clinical point of view the most important are serum $\alpha 1$ anti trypsin and $\alpha 2$ macroglobulin. The former inhibitor induces complete loss of activity, while binding to the latter results in approximately 20% loss of activity.

Trypsin is more toxic than chymotrypsin, particularly by rapid injection (Hendley *et al,* 1956) and is, therefore, only administered topically or by intramuscular injection.

3.2.2 Chymotrypsin

Chymotrypsin is prepared as a by-product during the fractionation of trypsin from an acid extract of mammalian, usually bovine, pancreas the fractionation isolates the enzyme mainly as the inactive pre-enzyme, chymotrypsinogen, which is activated by trypsin.

Chymotrypsin exists in a number of forms of which the two medicinally important are alpha and delta chymotrypsin. Alpha-chymotrypsin is converted to delta-chymotrypsin by standing in solution for several hours at high pH, and slow reconversion is possible on storage at pH 4.0 for months.

Like trypsin, chymotrypsin hydrolyses the peptide linkages in proteins and protein hydrolysates but with a preferential action upon bonds involving the carboxyl groups of l-amino acids.

The enzyme is inhibited by α_1-anti-trypsin and α_2-macroglobulin present in human serum, a specific α_1-anti-chymotrypsin has also been reported. The pancreatic trypsin inhibitor will also inhibit chymotrypsin.

Chymotrypsin has a lower general toxicity than trypsin; doses of up to 50 mg/kg on alternate days for 4 weeks failed to show pathological changes in the organs of experimental animals, and an increase in dose of up to 200 mg/kg body weight produced local necrosis, but no animal died.

Chymotrypsin will clot milk but not blood, in contrast to trypsin.

As with most proteinaceous enzymes, occasional anaphylactic reactions have been recorded.

3.2.3 Pancreatin

Pancreatin is a mixture of enzymes obtained by defatting and drying whole pancreas under carefully controlled conditions; by extraction of minced pancreas with chloroform water or acetone and drying *in vacuo;* or by alcohol fractionation of an aqueous extract.

The resulting product contains a variety of enzymes, of which the most important proteases are trypsin, chymotrypsin, carboxypeptidase and elastase. Pancreatin also contains amylase and lipase.

The ratio of enzyme activity and the total activity of the various enzyme components varies considerably according to animal source. Although pig and beef pancreatin are theoretically acceptable, only pig achieves a sufficiently high enzyme activity to be useful in medicine. Beef pancreatin is lower in tryptic activity and contains less amylase and little lipase in comparison to the porcine preparation.

3.2.4 Bromelains

The bromelains are a mixture of proteolytic enzymes prepared from the stem and fruit of the pineapple, *Ananas comosus* var. Cayenne.

The enzymes used in medicine are prepared by expressing the tissue of pineapple stems and precipitating the enzymes from the resulting juice with acetone (Heinicke and Gartner, 1957) or salts. Muruchi and Neurath (Muruchi and Neurath, 1960) tried unsuccessfully to produce crystalline enzymes from the crude extract, which possesses five proteolytically active components (El-Gharbawi and Whitaker, 1963).

Stem bromelain is active at neutral pH, catalysing the hydrolysis of a wide variety of proteins, peptides, esters and amides, especially at bonds involving basic amino acids, leucine or glycine. These include casein and haemoglobin, N-benzoyl-1-arginine ethyl ester, tosyl-1-arginine methyl ester, N-benzoyl-1-arginine amide, and benzoyl DL alanine ethyl ester.

General toxicity of bromelains is low. Chronic administration to rats for three months failed to show gross pathology in any organs studied (Moss *et al,* 1963). Gastro-intestinal disturbances rarely occur in man when the enzymes are administered orally.

3.2.5 Papain

Papain is prepared from the latex of the plant *Carica papaya.* It is activated in the presence of cysteine, which is generally included in the extraction and purification process. Salt fractionation with ammonium sulphate and sodium chloride followed by crystallisation and recrystallisation from alcohol is a common procedure for preparation of the medicinal enzyme.

Although pure papain has been prepared by chromatographic techniques (Skelton, 1968), the pharmaceutical preparation for use in medicine is usually a mixture of papain and chymopapain.

Papain experts its enzymic effects at an optimum pH around neutrality, hydrolyses peptides, amides and esters especially at bonds involving basic amino acids leucine or glycine. In this respect it resembles bromelains.

The papaya latex contains a natural activator; but papain may also be activated by hydrocyanic acid (Mendel and Blood, 1910) and many reducing agents.

Papain induces severe reactions in the lungs of dogs, rats and rabbits when administered as an aerosol or by instillation into the trachea; allergic reactions have followed repeated inhalations in man (Nava, 1969).

3.2.6 Chymopapain

Chymopapain is found in the latex of *Carica papaya* in association with papain. Its preparation in crystalline form was described by Jansen and Balls (Jansen and Balls, 1941), and its purification to a homogenous entity by Ebata and Yasunobu (Ebata and Yasunobu 1962).

Chymopapain hydrolyses peptides, amides and esters with a specificity similar to papain but exerts no activity on peptides, amides or esters containing aromatic amino acids.

3.2.7 Pepsin

Pepsin is a proteolytic enzyme of first attack, found in gastric mucosa. It is secreted as an inactive zymogen pepsinogen, which is converted to pepsin by autocatalysis. As expected from its source in the body, pepsin is optimally active at acid pH values, with a maximum between pH 1.6 and 1.8.

Pepsin used for medicinal purposes is normally of porcine origin, being prepared by acid aqueous extraction of minced mucosa. The resulting liquid is filtered as soon as possible, the enzyme is precipitated by ammonium sulphate fractionation, de-salted by dialysis and dried *in vacuo*.

The product thus obtained is a relatively crude pepsin, powder or translucent scales, having a slightly acid or salty taste. It is very hygroscopic, and undergoes slow auto-degradation in solution.

Northrop (Northrop, 1929-30) has described the preparation of a crystalline pepsin, which is rarely used for medicinal purposes. He demonstrated the inhomogeneity (Northrop, 1931) of the protein which was subsequently confirmed by Steinhardt (Steinhardt, 1939). At least four pepsin proteinases have been identified.

Pepsin catalyses the hydrolysis of almost all high molecular weight native proteins. Proteolysis is more effective if an aromatic ring is present in the side chain of the amino acid on either side of the split bond (Baker, 1951).

Specific pepsin inhibitors have been isolated from pepsinogen preparations and by repeated immunisation of animals. Pepsin is incompatible with pancreatin, tannin, concentrated ethanol and heavy metal salts.

3.2.8 Plasmin

Plasmin is the major fibrinolytic enzyme in blood. It normally circulates as the pre-enzyme, plasminogen, which is activated by a wide variety of proteases including trypsin, urokinase, streptokinase and specific tissue and blood activators.

In addition to its main effect of degrading fibrin to soluble peptides, commonly termed 'fibrin degradation products', plasmin also digests fibrinogen, anti-haemophilic factor (Factor VIII) prothrombin, and pro-accelerin (Factor V).

In many respects, plasmin is similar to trypsin and breaks essentially similar peptide bonds, having preferential specificity for peptides containing basic amino acids lysine and arginine.

It is also inhibited by similar inhibitors of which the most important medically are α_2-macroglobulin, α_1-anti-trypsin, and C_1-esterase inhibitor (Heimburger *et al,* 1971).

Plasmin for medicinal use is prepared by the activation of human plasminogen by highly purified streptokinase. The human plasminogen is isolated by alcohol and sephadex fractionation.

3.2.9 Streptokinase

This enzyme, as mentioned above, is a plasminokinase obtained from the culture filtrates of certain strains of *Haemolytic Streptococcus* group C. It has the property of activating plasminogen to form plasmin. Unlike trypsin and urokinase, streptokinase does not appear to have amino acid esterase or proteolytic activity, and may act by a two step process involving another plasma factor which is, at present, inseparable from plasminogen (Troll and Sherry, 1955) or plasminogen itself.

In medical use, streptokinase is occasionally combined with streptodornase, a streptococcal deoxyribonuclease which enzymically depolymerises deoxyribonucleic acid and deoxyribonucleoprotein. This increases the spectrum of activity for topical application.

3.2.10 Urokinase

Urokinase is another activator of plasminogen to plasmin. It is an esterolytic and proteolytic enzyme secreted in the human kidney and present in urine. Its proteolytic activity is confined to plasminogen only as a substrate and it is without action on fibrin, casein, or haemoglobin. Urokinase will, however, split specific amino acid ester substrates. Unlike streptokinase, urokinase, activates plasminogen directly in a single step reaction.

The enzyme was purified and crystallised by White in 1966 (White *et al,* 1966), and more recent work (Lesurk *et al,* 1967; Doleschel, 1975) has shown that at least three different forms of urokinase exist although these may be polymeric forms of a single enzyme.

Urokinase, like trypsin, may be inhibited under appropriate conditions by α_1-anti-trypsin (Clemmensen and Christensen, 1976) and α_2-macroglobulin (Ogston *et al,* 1973).

The enzyme has a very low toxicity in animal studies *(Pharmaceutical Enzymes,* Ruyssen and Lauwers, 1978).

3.2.11 Kallikreins

These are enzymes of very precise specificity isolated from a variety of tissues, and differing in physicochemical and enzymatic properties according to anatomical source. They exist as inactive pre-enzymes, pre-kallikreins, and are activated by trypsin or proteases, usually those liberated by cellular damage.

Kallikreins are involved in the inflammatory response of tissue to traumatic damage in that they liberate kinin peptides from kininogens. These kinins produce vasodilation, increase capillary permeability, produce pain and increase leucocyte migration (Lewis, 1963). Kinins also have an affect on smooth muscle including the small intestine and uterus.

Kallikreins used in medicine are prepared from hog pancreas, usually by organic solvent fractionation, followed by purification on DEAE cellulose then by column chromatography. A common contaminant of kallikreins is sialic acid which is removed using the enzyme sialidase.

Two iso-enzymes exist, normally identified as kallikreins A and B.

3.2.12 Ficin

Ficin is a proteolytic enzyme isolated from the sap of the South American fig tree, *Ficus glabrata* or *Ficus laurifolia.* The enzyme has a pH of optimum activity of 5, and is activated by reducing agents such as cysteine or hydrogen sulphide. Oxidation results in almost complete inactivation.

3.2.13 Sutilains

Sutilains is a mixture of proteolytic enzymes prepared by a fermentation process from selected strains of *Bacillus subtilis*. It is separated from the liquid culture medium by salt fractionation. Further purification may be achieved by adsorption using DEAE cellulose, further ammonium sulphate precipitation and fractionation on a CM cellulose column (McConn *et al,* 1964).

The enzyme is highly active in degrading both casein and haemoglobin substrates, its comparative activity on casein being approximately five times that of papain and 6-7 times that of trypsin and chymotrypsin.

The pH of optimum activity is between 6.5 and 8.0. The pH of optimum stability is slightly higher, being pH 6.5 to 9.7. Approximately 80% of the initial activity remained after 24 hours incubation at pH 7.0. Calcium ions acted as a stabiliser at every pH tested (McConn *et al.*, 1964).

The enzyme is inhibited by chelating agents and heavy metals. It contains 1g atom of zinc per mole of enzyme.

The commercial enzyme is probably composed mainly of subtilopeptidase A, an enzyme of very wide specificity. One enzyme isolated from *B. subtilis* BPN_1 strain attacks about one third of the peptide bonds in casein and a quarter of those in gelatin; ester bonds are also attacked but amide bonds are more resistant.

3.2.14 Brinolase

Brinolase is a fibrinolytic enzyme obtained from *Aspergillus oryzae* by controlled fermentation. It catalyses the hydrolysis of proteins and peptides especially at bonds involving the carboxyl groups of L-arginine, L-leucine, or L-glutamate, and will activate both trypsinogen and chymotrypsinogen.

3.3 THE FORMULATION OF PROTEOLYTIC ENZYMES FOR MEDICINAL USE

The general principles of the formulation of any drug substance for maximum pharmacological effect naturally apply to the preparation of proteolytic enzymes for medical use.

The enzyme must be so compounded as to reach the patient in a fully active form, be released at the site at which it is required to exert its effect; be uniform in potency and biological activity and sufficiently stable to withstand normal storage conditions for a reasonable period. It must be compatible with the excipients and other ingredients of the medicinal composition, and not react with its containers or their closures.

By their nature, many enzymes are sensitive to heat, oxygen, heavy metals, inhibitory substances; light, changes in pH and, in the case of proteolytic enzymes, they undergo autolysis. These adverse effects are enhanced in the presence of moisture. It is, therefore, essential that proteolytic enzyme preparations for medicinal use are presented, as far as possible, in the dry state for maximum stability. Hence strict moisture limits are necessary for tablets, powders and granules. Ointments and topical preparations are formulated in anhydrous bases, and injections are generally sterile freeze-dried powders ready for reconstitution with a suitable diluent immediately before use, or as suspensions in oil. Because of the inconvenience of lyophilised injections and the pain associated with the injection of oils, stabilised aqueous solutions are used where possible, but inevitably shelf life is reduced.

Bacterial decomposition of the enzyme protein structure is an important feature to consider, and proteolytic enzyme preparations are either presented sterile or with stringent limits on their bacterial population.

Although enzymes have a pH of optimum activity, most also have a pH of optimum stability. This fact is taken into account during pharmaceutical formulation. Unfortunately, in many cases, the pH of optimum stability is far removed from the pH of optimum activity, and final application of the enzyme may involve the use of buffer solutions to bring the enzyme from a pH of best stability to that of optimum activity.

The effects of moisture and other adverse conditions can be minimised by the use of stabilisers, calcium for example in the case of trypsin and chymotrypsin. Cysteine stabilises papain and ficin against oxidation.

Deleterious heavy metal ions are excluded during manufacture, all processes being conducted in stainless steel, glass, or glass-lined vessels and equipment. Natural inhibitors occurring in the tissues of origin are removed during the purification process.

Finally, because of their protein structure, proteolytic enzymes for oral administration need protection from acid pepsin present in the stomach. Tablets and granules, therefore, are usually protected by enteric coatings resistant to gastric juice but dissolving in the higher pH of the small intestine and releasing the dose of enzyme at this point.

Use of more advanced technology suggests that by encapsulation or immobilisation on inert support systems proteolytic enzymes can be more effectively transported in a fully active state to their site of action. This principle has been used by Gonda (Gonda, 1977) for pancreatic enzymes, supported on carboxymethyl cellulose or agarose.

The use of liposomes also shows promise for the transport of encapsulated enzymes to specific sites for target release.

3.4 STANDARDISATION

Most drugs are available as pure and discrete chemical entities, capable of precise chemical analysis and hence a dose definition on a weight basis, e.g. in milligrams, is a logical and valid proposition.

Biological drugs, on the other hand, are often complex proteins or mixtures of proteins whose chemical estimation is difficult on a routine basis, or meaningless in terms of their pharmacological effect. In these cases, it is customary to resort to biological or biochemical assay procedures which more closely define potency in terms of their expected biological effect.

Proteolytic enzymes fall into this class of drugs, since they rarely approach 100% purity, nor do chemical assay procedures give a simple relationship to clinical effect. The method of assessing potency must, therefore, be based on the observation of the rate of enzyme catalysis on a specified substrate under standard conditions.

Although a number of proteolytic enzymes have appeared in the various national pharmacopoeias over the years, national and international agreement on methods of assay and on units has not yet been achieved. Many different units exist, which make difficulties in the interpretation of clinical results and dosages of different or even the same enzyme product.

On the other hand, it is essential for the clinician using proteolytic enzymes to be presented with products of reliable, consistent and predictable performance.

Early attempts at standardisation mainly involved the use of natural protein substrates such as egg albumin or casein and were primarily limit tests for total proteolytic activity. Although minimum potency was defined, the exact activity was not fixed and considerable inter-batch variation was possible.

Later, mainly owing to the efforts of individual research workers such as Anson (Anson, 1939), precise proteolytic activity on natural substrates such as haemoglobin was established and a unit definition was possible. In the case of haemoglobin, these units were obtained by comparison of split products produced in defined time against a chemically definable standard tyrosine, the split products and tyrosine, both giving a measurable blue colour with alkaline Folin-Ciocalteu reagent.

More recent attempts to define enzyme activity of proteolytic enzymes has concentrated on their esterase action on specific linkages of well defined chemical substrates; for example, benzoyl l-arginine ethyl ester for trypsin (Neurath and Schwert, 1950) and enzymes of similar specificty, acetyl-l-tyrosine ethyl ester for chymotrypsin, and carbobenzoxy-l-tyrosine for pepsin.

This principle was adopted by the United States National Formulary (*United States National Formulary* XIV, 1975) for pharmaceutical grades of trypsin and chymotrypsin for medicinal use, although the units defined were arbitrarily based on those in current use, mainly the 'Armour Unit', (Christie, 1960). (*United States Pharmacopoeia* XIX, 1975).

The first attempt to rationalise the standardisation of enzymes for medical use, on an international basis, apart from the limited activities of the World Health Organisation, was undertaken by the International Federation of Pharmacy.

The International Commission for Pharmaceutical Enzymes of the above body adopted the unit principle, first proposed by the International Union of Biochemistry, that a unit of activity should be that amount of enzyme that would catalyse the transformation of one micromole or one micro-equivalent of substrate per minute under defined conditions, usually at 25°C. Methods of assay involving this principle of unit standardisation have now been proposed for most medicinally important enzymes (*Pharmaceutical Enzymes,* Ruyssen and Lauwers, 1978). More recently, standardisation in catals has been proposed, but this results in a unit of inconvenient size and complex conversion to the former system.

Unfortunately, this principle is difficult to apply to mixed proteases often used for medicinal purposes, since the synthetic substrates used are rarely entirely specific, and the resulting overlap of activity leads to artificially high total results. In such cases, there is a continuing place for protein substrates such as casein and haemoglobin, particularly if the rate of formation of split products can be expressed in terms of micromoles per minute. In the latter case, the desirable standard definition of a unit can still be applied.

3.5 MEDICINAL USE

It is proposed to describe the medicinal use of the various proteolytic enzymes discussed previously in this chapter under broad therapeutic readings.

3.5.1 Aids to digestion and use in digestive enzyme deficiency states

3.5.1.1 AIDS TO DIGESTION

Traditionally, this is the most logical and earliest use of proteolytic enzymes in more recent history. Many of the first enzymes discovered were extracted from organs of the animal digestive system, and these crude extracts were used as 'digestive aids'. Early application was particularly non-scientific, mixtures of pepsin and pancreatin which are mutually incompatible being common. The B.P.C. 1934 listed no fewer than ten elixirs, mixtures, solutions and powders containing pepsin.

Pepsin may be administered in acid solution to increase the digestive power of gastric juice. Such therapy would appear logical where there is a genuine deficiency of pepsin secretion, or for partial gastrectomy, but good objective evidence of its effectiveness even in such conditions is lacking. Probably for this reason the use of pepsin as a digestive aid is declining.

More solid evidence is available for the provision of enzyme supplements of animal, vegetable or microbiological origin to aid pancreatic digestion.

In a recent publication (Di Magno *et al* 1977) six patients with proven pancreatic insufficiency were given eight tablets of pancreatic extract with meals or two tablets hourly. The enzyme activity of aspirated pancreatic juice was followed over a period up to three hours following initial dosing. Tryptic activity was significantly elevated, particularly after eight tablets with the meal. There were, however, very substantial losses of activity. Only 22%, or 16%, of the trypsin administered respectively by the two schedules survived to the duodenal aspiration site at the ligament of Treitz. The authors attribute these losses in potency to intragastric and intraduodenal acidity which is generally increased in pancreatitis compared to the healthy subject. At no time by either schedule did the tryptic activity of the pancreatic juice approach that of the healthy subjects (See Table 1). In spite of this, faecal nitrogen estimations proved that azotorrhea was corrected in five out of the six patients by eight tablets with meals.

Better recovery of proteolytic activity may be achieved by protecting the enzymes against gastric acid by the use of enteric coatings such as cellulose acetate phthallate/diethyl phthallate combinations or waxes. Concomitant administration of antacids or gastric acid secretion inhibitors such as cimetidine or metiamide have also been suggested (Regan *et al,* 1976; Saunders *et al,* 1977).

Alternative approaches have been to combine the pancreatic enzyme content with proteases of vegetable or fungal origin which have better stability in an acid medium.

Twelve adults with chronic pancreatic deficiency were treated with a dual tablet containing pancreatin plus ox-bile extract in an enteric coated central core, surrounded by compressed bromelains, (Knill-Jones *et al,* 1970). These bromelains are active between pH 3 and 8 and are not inactivated below pH 2. The authors measured faecal fat content as a criterion for effectiveness and proved that these combined tablets reduced the amount of faecal fat more than simple pancreatin tablets. Strangely, removal of the bromelain outer coat reduced the effectiveness of the tablets, suggesting that the bromelains had contributed to the lipolytic effectiveness of the combined tablet.

Table 3.1

Postprandial Enzyme Concentrations in Duodenal Aspirate of Six Healthy Subjects and Six Patients with Pancreatitis (Sequential Samples).†

Group	Enzyme (U/ml. of aspirate) postprandially					
	AT 30 MIN	AT 60 MIN	AT 90 MIN	AT 120 MIN	AT 150 MIN	AT 180 MIN
Trypsin:						
Healthy subjects	114 ±18	118 ±17	144 ±28	150 ±26	173 ±28	132 ±30
Patients:						
Without treatment	0.7 ±0.3	0.3 ±0.4	0.3 ±0.1	0.3 ±0.1	0.2 ±0.1	0.5 ±0.3
With pancreatin:						
8 tablets with meal	0.8 ±0.2	3.3 ±1.3	4.1 ±1.1	4.0 ±1.8	1.4 ±0.9	1.2 ±0.9
2 tablets/hr.	2.0 ±1.3	2.0 ±0.5	2.1 ±0.8	2.1 ±1.0	1.6 ±0.6	2.0 ±1.2
Lipase:						
Healthy subjects	522 ±126	433 ±128	459 ±106	464 ±105	532 ±125	397 ±115
Patients:						
Without treatment	0.4 ±0.1	0.4 ±0.1	0.5 ±0.2	0.5 ±0.1	0.5 ±0.2	0.4 ±0.1
With pancreatin:						
8 tablets with meal	0.8 ±0.2	3.2 ±1.6	3.2 ±1.0	4.7 ±3.7	1.3 ±0.6	0.7 ±0.3
2 tablets/hr.	0.9 ±0.3	2.0 ±0.7	2.4 ±1.4	2.1 ±1.1	1.1 ±0.6	2.0 ±1.1

† All values means ± SEM; those in pancreatitis values in health (P 0.001); in pancreatitis, all values with pancreatin (except trypsin at 30 min) those without, but with no difference between effects of treatment schedules.

Reprinted, by permission, from the *New England Journal of Medicine* **296**, 1320 (Table 2 on that page) (1977).

In a larger double blind trial, 50 patients with a variety of gastro-intestinal disorders were treated with either a combined tablet composed of fungal proteases, cellulases, pancreatin and ox-bile extract, or a placebo. Dyspeptic symptoms were relieved more effectively in the active group (Karani *et al,* 1971).

One problem faced by physicians using pancreatic preparations currently available is the wide variation in potency, a situation further aggravated by the variety of units in which the potency is expressed. Drummond and co-workers (Drummond *et al,* 1977) analysed twenty commercial pancreatic extract products using the F.I.P. methodology (*Pharmaceutical Enzymes,* Ruyssen and Lauwers, 1978), to try and guage the extent of this variation in potency using a standard assay technique.

3.5.1.2 CYSTIC FIBROSIS

A very well substantiated indication for pancreatic enzyme replacement therapy is cystic fibrosis. The majority of children suffering from this condition require regular doses of high potency pancreatin preparations because of continued pancreatic insufficiency and consequent malabsorption of proteins and fats. The high potency pancreatin preparations are available as tablets, capsules, granules, or powders. Originally, enteric coating was used to provide better stability in gastric juice. Unfortunately, such coatings are designed to dissolve in alkaline media, and the pH of duodenal juice of cystic fibrosis patients is frequently neutral or slightly acid. Enteric coatings have, therefore, been either drastically reduced in thickness or omitted. Trials with pancreatin preparations have shown that even in high doses control of the pancreatic insufficiency, although improved, is not complete (Harris *et al,* 1955; Goodchild *et al,* 1974). Further research into improved methods of delivering the pancreatic extract to provide maximum activity in the duodenum is, therefore, indicated.

3.5.2 Topical debridement

The removal of purulent exudate, necrotic tissue or hard and resistant eschar is a vital prerequisite for the effective cleaning and sterilisation of burns, ulcers, traumatic and surgical wounds, also for the initiation of the healing process. Similarly in the treatment of chronic bronchial disease, the liquefaction and drainage of viscous and often infected sputum is a valuable adjunct to treatment. Such cleansing processes can be achieved mechanically, but the techniques are tedious, and often cause suffering to the patient and extend the damage already evident.

As might be expected, numerous reports can be found in the literature discussing the usefulness of proteolytic enzymes in liquefying and removing necrotic tissue. Because of the complexity of the substrates, for example fibrin, mucoproteins, collagen, a bewildering variety of topical enzyme applications have been suggested.

The picture is further complicated by the fact that many authors have combined their proteolytic enzyme applications with antiseptics or antibiotics,which renders unequivocal interpretation of their results difficult. Generally, enzymes are best applied in solution as wet dressings; if used inpowder form the wound surface must be moist.

Stability of the enzyme preparations used in a hostile environment must be suspect since they are often subjected to higher than optimum temperature, beset with natural inhibitors, and are in contact with micro-organisms, for long periods. Attempts to offset this effect have been made by the addition of stabilisers to solutions (e.g. calcium in the case of trypsin) or the use of non-aqueous ointment bases which are water miscible and slowly release their proteolytic activity in contact with body fluids. Such ointments are not of great value for the removal of dry eschar.

3.5.2.1 WOUND DEBRIDEMENT *IN-VITRO* AND CLINICAL EXPERIENCE

Some of the earliest *in vitro* experimental work on the use of proteolytic enzymes in wound debridement was reported by Reiser, Patton and Curtis (Reiser *et al,* 1951) who extended the laboratory examination of the effects of trypsin on fibrin and blood clots to a number of isolated tissues and were able to demonstrate that necrotic muscle was rapidly attacked and gangrenous and normal skin, fascia and blood vessels were partially digested.

Later work by this group (Reiser *et al,* 1953) compared the effects of chymotrypsin with trypsin and concluded that although the lysis of fibrin by chymotrypsin is only 25% of that by trypsin alone, a combination of the two enzymes shows enhanced lytic activity to 270% of that of trypsin alone.

Reiser and his co-workers then applied their *in vitro* findings on trypsin to the treatment of a number of purulent or necrotic lesions in patients including haematoma, osteomyelitis, and infected amputation stumps (Reiser *et al,* 1951).

It is common experience that delayed treatment of wounds with topical antibiotics is less effective than immediate treatment following injury. It was postulated by Rodeheaver and his co-workers (Rodeheaver *et al,* 1974) that this situation could be improved by the topical treatment of the wounds with proteolytic enzymes prior to local antibiotic therapy. They tested their hypothesis with an elegant series of experiments with standardised injuries in guinea pigs, followed by experimental infection of the wounds with *Staphylococcus aureus.* After a delay of four days, the wounds were treated with varying concentrations of trypsin solution from 2 500 to 50 000 U.S.N.F. units in 0.1 ml with and without the addition of benzyl penicillin at varying intervals.

Inflammatory response and viable bacterial counts from the wounds showed that enzymatic hydrolysis of the wound coagulum significantly enhanced the value of the antibiotic in the delayed treatment of contaminated wounds. The effectiveness of the trypsin was increased by prolonging its application or by repeated application. The enhancement reported was applicable to both topically applied and systemic antibiotic, and the authors noted that trypsin itself showed measurable antibacterial activity against *Staphylococcus aureus.*

These animal studies were a confirmation under controlled laboratory conditions of the experience of Miller *et al* (Miller *et al,* 1953) who successfully treated nineteen patients with wound infections. Application of trypsin was by solution in Sørensen's buffer and subsequent incorporation into a water soluble ointment base. The trypsin changed thick and viscid exudate into a fluid material which could easily be drained. With concomitant antibiotics the infections were cleared and the wounds healed.

Combined treatment of trypsin and chymotrypsin either in solution as a dry powder in a capsule proved 200% more rapid than an equal concentration of trypsin alone, in a variety of infected injuries including gas gangrene of the leg and decubitis ulcers (Reiser *et al,* 1953). This principle was used by other physicians later (Irgang, 1960; Spencer, 1967) in the treatment of pyodermas with a combined trypsin-chymotrypsin ointment. The authors were particularly impressed with the prompt removal of crusts and exudate covering the infected areas (Spencer, 1967).

The presence of desoxyribonucleoprotein in infected inflammatory exudate led to the search for enzymes with a broader spectrum of activity than trypsin or chymotrypsin, which would depolymerise fibrin and also liquefy the desoxyribonucleoprotein. Such an enzyme system is streptokinase-streptodornase, enzymes isolated from filtrates of Lancefield group C beta haemolytic streptococcal culture medium. This enzyme combination was very effective in cleansing deep seated war wounds with multiple sinuses, although trypsin was more effective in the treatment of osteomyelitis and in cleansing surface wounds (Spittler and Parmenter, 1954).

The efficacy of the local use of trypsin in the treatment of bone infections was also reported from a Hamburg hospital. Trypsin plus antibiotic was introduced into the medullary cavity to act directly upon the purulent material which was effectively isolated from systemically administered antibiotics by lack of blood supply. Considerable success was reported in a number of cases (Profitos, 1965).

Similar results using streptokinase/streptodornase have been reported by other authors (Miller *et al,* 1953).

3.5.2.2 BURNS

Early removal of necrotic tissue and debris from severe burns is an important preliminary to skin grafting. A clean surface is essential to successful grafting. This is often achieved by surgical or mechanical means. Even in those burns where grafting is not required, freedom from dead tissue and eschar produces a cleaner wound, less likely to be extensively colonised by bacteria and in which tissue repair is likely to proceed smoothly.

The application of trypsin in extensive burns by Stucke (Stucke, 1949; Stucke, 1954) showed that in trypsin treated cases the time required for complete healing was more than halved and keloid formation was reduced. The percentage of patients requiring skin grafts or subsequent plastic surgery was also substantially less in the trypsin treated series. Percentage of takes of skin grafts on trypsin cleaned burned areas was likewise much improved (Wilde and Derry, 1953).

Treatment of children by Trendtel (Trendtel, 1950) and Leuterer, (Leuterer 1957) confirmed the improvement in healing of second and third degree burns and scalds when trypsin was applied. Miller and others reported successful treatment of severely burnt military casualties using trypsin compresses (Miller *et al,* 1953).

In most cases, the topically applied trypsin was accompanied by local antibiotic and conventional support therapy. Trypsin is applied as a wet dressing in buffer solution which requires frequent replacement to sustain the proteolytic effect. Stucke, however, suggested a novel trypsin-antibiotic dressing which was stored dry and moistened immediately before use; when covered with a water impervious dressing combination it could be left in place for a longer period.

In spite of the success reported in the 1950s, the use of trypsin as a biochemical debriding agent for burns has not attained widespread acceptance. Around the same time, other investigators (Howes *et al,* 1959; Burke and Golden, 1958) were utilising enzymes of bacterial and fungal origin and papain. In particular, it was anticipated that collagenases/proteinases isolated by the ammonium sulphate and methanol fractionation of *Clostridium histolyticum* H-4 would be superior burn debriding agents. Early promise (Howes *et al,* 1959; Hamit and Upjohn, 1958) was not upheld by later work (Hamit and Upjohn, 1960).

Interest in enzymic debridement was revived by a series of *in vitro* experiments comparing the effectiveness of various proteolytic enzymes in digesting isolated burn eschar (Prytz *et al,* 1966; Silverstein *et al,* 1973) and in animal burn models (Krizek *et al,* 1974; Zawacki, 1975).

Comparative digestion experiments (Silverstein *et al,* 1973) were carried out using bromelain, sutilains, and *Clostridium histolyticum* collagenase with either split-thickness pig skin or isolated human eschar as substrates. The amount of leucine produced gave a measure of digestion rate. In addition to the natural skin products, purified pig skin gelatin and microcrystalline collagen were also tried as substrates for comparative tests. Separate digestion experiments were supplemented by sequential addition of the enzymes to determine synergistic effects.

Overall, sutilains was the most effective of the enzymes against pig skin and eschar, showing both proteolytic and some collagenolytic activity, an important attribute for *in vivo* burn debridement.

Confirmation (Krizek *et al,* 1974) of this was sought in rats and pigs, with experimental full thickness scald burns subsequently inoculated with different strains of *Pseudomonas aeruginosa.* A total of 190 rats, divided into groups, and a pig had their infected burns treated with various combinations of treatment schedules including sutilains ointment alone, or combined with the antibacterial agents mafemide acetate, 10% and silver sulphadiazine, 1%. One disturbing result of this study was that the treatment with sutilains ointment seemed to predispose burn wounds to massive sepsis, lethal in the animals concerned. Unfortunately, the incorporation of the antibacterial agents did not protect the animals against the sepsis and subsequent death, and the authors warned caution in the clinical use of sutilains debridement in contaminated burn wounds. Zawacki (Zawacki, 1975) reached similar conclusions following experimental work with heat injured guinea pigs.

In spite of this cautionary note, Pennisi and co-authors (Pennisi *et al,* 1975) compared the progress of 50 moderate to severely burned patients treated with silver sulphadiazine ointment alone with 62 similar patients in which the silver sulphadiazine ointment was supplemented by debridement with sutilains ointment. They concluded that skin grafting could be undertaken earlier and discharge of the of the patients achieved approximately 17 days sooner when enzyme debridement was used. Furthermore, mortality rate in the unit was reduced, and on this basis they refuted the suggestion that proper enzymatic debridement with sutilains ointment increases the incidence of wound sepsis.

Additional substantiation of this claim has been made in two recent papers (Levick *et al,* 1977; Dimick, 1977). Sutilains ointment was effective in removing eschar and necrotic tissue and permitted early skin grafting, but patient mortality was not significantly affected.

3.5.2.3 ULCERS

Irrespective of fundamental origin and cause, thorough cleaning of ulcers with removal of infected material and necrotic debris is an essential preliminary to granulation and epidermal overgrowth.

As might be expected, debridement with proteolytic enzymes has been proposed to achieve rapid non-traumatic removal of unwanted proteinaceous material from ulcers. *In-vitro* comparison of streptokinase-streptodornase and trypsin at body temperature on necrotic crust, fibrin, and purulent exudate removed from leg ulcers indicated the superiority of streptokinase-streptodornase, although no difference could be demonstrated between the two enzymes on 3-day old blood clots removed from the same ulcers (Hellgren and Vincent, 1977).

Clinical confirmation of the effectiveness of streptokinase-streptodornase was obtained in 15 cases forming part of a mixed study of enzyme cleaning of various surgical conditions. As an adjunctive therapy, the enzyme debridement performed a useful role and toxic reactions were minimal (Wright *et al,* 1954).

Another comparative study (Prueter *et al,* 1957) on 15 patients using chymotrypsin on one ulcerated leg and a commercially available bacterial protease on the other, showed a marked superiority of chymotrypsin in 12 of the patients. In a single patient, chymotrypsin was also more effective than trypsin. The enzymes were applied to the decubital or necrotic type ulcers either as a wet gauze dressing or in a methyl cellulose gel.

Trypsin has also improved the healing of infected cutaneous ulcers when applied as a solution in saline or Sørensen's phosphate buffer on a gauze pad. Solution should be allowed to act on the necrotic tissue for three hours, and the application should be made several times daily. If the lesion is appropriately sited, dry powder applications of trypsin may be made.

The greater effectiveness of chymotrypsin under specified conditions has led to the formulation of an ointment containing the two enzymes mixed as a high potency concentrate in a polyethylene glycol base. This ointment, as tested and used clinically, also contains hydrocortisone acetate and neomycin palmitate. In 1960 a report on the treatment with this ointment of 251 patients suffering from a variety of dermatological ailments showed improvement or a complete cure of 175 patients, the majority being leg ulcers (Cornbleet and Chesrow, 1960).

This early investigation was uncontrolled, and more recent confirmation of the effectiveness of this combination in the treatment of leg ulcers has been published by Gordon (Gordon, 1975). In a double blind experiment on 19 patients with post-thrombotic leg ulcers he showed a faster healing rate in the group treated with the ointment containing trypsin-chymotrypsin compared to the group treated with hydrocortisone and neomycin alone.

Sutilains ointment has also been widely used for treating stasis ulcers, but the work is largely uncontrolled and the results described subjective.

Skin ulcers very refractory to treatment have been treated using terramycin-plasmin solution as a wet dressing. Results were most striking with multiple ulcers (Spier and Clifton, 1954).

3.5.2.4 CHEST DISEASE

Thick, tenacious sputum which impedes respiration is a characteristic of bronchial asthma, chronic bronchitis and other respiratory conditions. In acute and chronic chest infection, adequate drainage of infected exudate from the pleural cavity improves penetration and increases the efficacy of antibiotics, allows re-expansion of the lungs, and facilitates surgery where necessary. The secretions in question are inaccessible, and their liquefaction and removal presents a problem. Although some rather unsuccessful use of papain aerosol was attempted in the late 1930s, it was not until 1949 that the successful use of enzymes was reported. The enzyme used was streptokinase, and successful liquefaction of fibrinous, purulent and sanguinous pleural effusions was achieved (Tillett and Sherry, 1949).

Over the following two years, this initial work was substantiated (Tillett and Sherry, 1951). In particular it was shown that in addition to liquefaction of empyema, traumatic and post-operative haemothorax could be drained without surgical intervention (Read and Berry, 1950). Sufficient streptokinase to give a concentration in the fluid of 100-500 units, was injected into the thoracic cavity and the liquefied material aspirated not more than 12 to 18 hours later.

There was growing concern over the use of streptokinase, since the body secretes no natural inhibitor to this enzyme, unlike trypsin or chymotrypsin. Some control of unbridled activity is provided in that streptokinase exerts its fibrinolytic effect by activation of plasminogen to plasmin. Well defined inhibitors of plasmin occur in blood, most body fluids, and exudates.

Notwithstanding, trypsin was subsequently tried as an alternative to streptokinase and has proved very successful. Just over 20 years ago, poliomyelitis was common and patients with the disease often required artificial respiration. One complication of this procedure is the accumulation of thick and tenacious secretions, which impede respiration and block tracheostomy tubes. Trypsin applied via a bronchoscope or aerosol spray through the tracheostomy tube, at a concentration of 125 000 units in 3 ml of Sørensen's phosphate buffer was remarkably effective in reducing the viscosity of these secretions and promoting their removal (Camarata, 1956; Kinnear-Wilson and Stevenson, 1957; Steigrin and Scott, 1952; Peck and Levin 1952).

In a variety of infected chest diseases, the direct instillation of trypsin solution by thoracentesis resulted in liquefaction of infected material which could be withdrawn by direct aspiration. Pretreatment of the patient with anti-histamine is recommended to reduce the incidence of trypsin sensitivity reactions. Some investigators reported a dramatic reduction in bacterial contamination of the thoracic cavity, including those infected with *Tubercle bacillus* (Roettig *et al,* 1952; Seiter, 1955).

Trypsin treatment appears less effective in the case of pure asthma; in one trial only 2 out of 17 cases improved (Prince *et al,* 1954).

The ancillary use of trypsin to clear tenacious secretions in preparation for bronchoscopy (Mohnke, 1953) and surgery (Habeeb *et al,* 1954) is also of value.

In spite of its effectiveness, trypsin is liable to give rise to troublesome side effects, including hoarseness, and soreness when administered by aerosol. Alternatives such as chymotrypsin (Costa and Fontana, 1959) and pancreatic dornase (Salomon *et al,* 1954) have also been tried by instillation or aerosol. Pancreatic dornase was particularly effective if the secretion was mainly pus. Dornase aerosol proved to be less irritating than trypsin and did not cause hoarseness.

For the treatment of tenacious sputum in chronic bronchitis, a further advance came with the use of proteolytic enzymes administered orally. The effect of a mixed trypsin and chymotrypsin tablet on thinning sputum was first described by Simon and Harman 1961 (Simon and Harman,1961) following work on oral chymotrypsin by Taub (Taub, 1960). Comparison was made between trypsin-chymotrypsin tablets, potassium iodide and ammonium chloride for thinning and expectoration of thick mucoid sputum in a series of male patients suffering from chronic lung disease. Changes in the volume and viscosity of 24-hour specimens of the patient's sputum were recorded. All the drugs decreased sputum viscosity, potassium iodide being very slightly more effective than oral proteolytic enzymes at the dose of enzyme used; both were more effective than ammonium chloride. The technique of measuring the viscosity of the sputum depended on blending the samples with water, then measuring the viscosity of the diluted sample in an Ostwald U-tube viscometer. It is possible that errors could have been introduced by this procedure since sputum is a non-Newtonian fluid and its structure and viscosity properties are altered by shearing. At about the same time, Bruce and Quinton, working with a specially designed viscometer, attempted to assess the effects of mixed trypsin-chymotrypsin tablets administered orally on a double blind basis (Bruce and Quinton, 1962). Their use of oral enzymes was stimulated by previous successful use of chymotrypsin injected intramuscularly in such conditions as bronchial asthma, bronchitis and sinusitis (Parsons, 1958; Taub, 1960; Billow *et al,* 1960). Nineteen patients suffering from chronic bronchitis were given dummy tablets and a similar group of 21 patients recieved active proteolyticenzymes as enteric coated tablets. An initial sample of sputum was taken from each patient and at weekly intervals during the four weeks of the trial. Sputum viscosity was measured on a spring and plate viscometer at constant rotation. Relative viscosity was indicated in arbitrary units on a scale by pointer deflection. The results were subjected to statistical analysis and indicated a marked reduction in viscosity or liquefaction of sputum in those patients receiving the oral proteolytic enzymes.

The results of Bruce and Quinton were confirmed by later authors using subjective assessments of the sputum viscosity and appearance (Bouvier *et al,* 1962) or objective measurement with a Ferrier viscometer (Bourgois and Bourgois, 1964). A reduction in viscosity of 43.8% was recorded in the latter case, following oral trypsin-chymotrypsin therapy.

The exact mode of action of the enzymes given by mouth is obscure, since by traditional therories of protein absorption they should not pass through the intestinal barrier into the bloodstream, and even if absorption does occur, massive amounts of anti-proteases are present in the blood. Nevertheless, the clinical evidence is convincing, and it is possible that an indirect mechanism is involved. This subject is discussed more fully under the anti-inflammatory action of proteolytic enzymes, Section 3.8.2.

3.5.2.5 ACNE AND SKIN CLEANSING

Topical preparations for the cure or relief of acne form an interesting boundary area between medicine and cosmetics, particularly if the therapy incorporates an element of deep skin cleansing.

Several authors have incorporated the topical treatment of acne with trypsin and chymotrypsin within general clinical reports on the use of these enzymes in dermatology (Cornbleet and Chesrow, 1960; Jenkins 1969; Ruggiero, 1967).

Cayle (Cayle 1963) in a review of 'The potential of enzymes for topical application', highlighted interesting possibilities for cosmetic uses of proteases.

Few attempts at formulation and practical use have been reported. A beauty mask containing trypsin, chymotrypsin and lipase was described separately by authors in Czechoslavakia (Fertek *et al,* 1968) and Italy (Mantegazza and Rovesti, 1958).

Most of the experimental preparations described by preceding investigators have the disadvantage of instability, required refrigeration for storage and exhibited a very limited shelf life. Christie and Chudzikowski (Christie and Chudzikowski, 1971) made the first serious attempt to formulate a biological mask in an inert base which was stored dry and reconstituted with warm water immediately before use. The proteolytic enzymes used were a concentrate from pancreas, being a combination of trypsin and chymotrypsin in the approximate ratio of 6:1. A number of different carriers were tried based on Guar gum and Carbitol, together with buffers and thickening agents. An acceptable powder was developed which showed good stability after up to 5 months' storage, and which retained approximately 80% of its activity for the 20-30 minutes' skin contact time after reconstitution. Subjective trials on six volunteers showed marked improvement in all cases at the end of a series of weekly applications. No untoward effects were observed. Improvement was assessed on the basis of the degree of inflammation and redness of spots and the texture and cleanliness of the skin on the face. This was regarded as a preliminary trial only and no placebo application was used in parallel. Unfortunately, no follow up to this approach to skin cleansing has been reported.

3.6 SOME OTHER AREAS OF APPLICATION

3.6.1 Ophthalmology

In the surgical treatment of cataract, removal of the opaque or partially opaque lens or lens capsule is essential. Difficulty may be experienced in freeing the lens from its supporting structures, particularly the zonule. In cases of zonular resistance damage to other eye structures may result.

Early in 1957 Barraquer, working in Barcelona injected a solution of chymotrypsin into an eye which had suffered an injury two years earlier and in which a substantial amount of blood had remained in the vitreous. He hoped that the emzyme would digest the coagulated blood and thus assist in its reabsorption. Examination of the eye subsequent to the chymotrypsin treatment revealed that the lens had been completely freed by the action of the chymotrypsin. This accidental discovery was followed by a series of animal and eventually human trials to establish weak solutions of chymotrypsin as a valuable adjunct to cataract surgery.

Barraquer concluded that a 1:5000 solution of chymotrypsin introduced into the anterior chamber of the eye was sufficient to dissolve the zonule and free the lens for conventional intracapsular extraction techniques. Risk of retinal detachment was reduced because the traction required for mechanical rupture of the zonule is avoided (Barraquer, 1958a,b; Barraquer, 1959).

Barraquer's work was confirmed in the United Kingdom by Cogan (Cogan *et al,* 1959), in the U.S.A. by Jenkins (Jenkins, 1960), in Belgium by Claes and Legrand (Claes and Legrand, 1958), and in Norway by Boberg-Ans and Badsberg (Boberg-Ans and Badsberg, 1959). As a result, enzymatic zonulolysis is accepted as a standard technique for the removal of cataracts in patients of twenty years of age and upwards. After the age of fifty, use of chymotrypsin is less necessary because the zonule becomes degenerate; however, the use of the enzyme ensures that all zonular resistance is eradicated.

The use of chymotrypsin does not raise intraocular pressure in glaucomatous eyes (Gambos and Oliver, 1967), and even the accidental use of 200 times the recommended concentration of chymotrypsin did not result in complications (Ray, 1964). Incompatibility with penicillin by direct admixture in the anterior chamber is reported (Koke, 1967); the activity of the chymotrypsin is reduced, hence the antibiotic must be introduced after the chymotrypsin has disrupted the zonular fibres.

Enzymatic zonulolysis has been used in veterinary surgery as an aid to cataract surgery in dogs (Startup, 1960; Startup, 1967). Higher concentrations of enzyme are needed to digest the zonular fibres, together with mechanical tension, because the zonular fibres are surrounded by protective acid mucopolysaccharides.

Ligneous conjunctivitis is a rare disease, characterised by thickening of the conjunctival membrane, the formation of a pseudomembrane with loss of flexibility, and development of a pseudotumourlike appearance. In many cases, the cornea of the eye may be involved and in these cases there is a severe risk of the loss of the eye. Successful treatment of this condition with a combination of hyaluronidase and chymotrypsin has been published (Francois and Victoria-Trancosa, 1968; Firat and Tinaztepe, 1970).

More recently, chymotrypsin was used with some success to treat the vitreous haemorrhages of patients with sickle-cell anaemia.

3.6.2 Otolaryngology

Analysis of various types of middle ear effusions has shown that deoxyribonucleic acid and protein bound carbohydrate are the constituents most likely to be responsible for the high viscosity which renders the removal of such effusions very difficult. Purulent effusions are highest in deoxyribonucleic acid content, but observations of relative proportions of the various components in different types of effusions have led to the conclusion that protein bound carbohydrate is mainly responsible for the high viscosity. It is probable, therefore, that disruption of the protein linkages could reduce viscosity to such a degree that the effusion could be removed or re-sorbed. Gessert and his co-workers (Gessert *et al,* 1960) decided to test this hypotheses *in vitro.* Various effusions removed from human middle ears were treated with trypsin, chymotrypsin, pepsin, papain, pancreatic and bacterial dornase, hyaluronidase, lysozyme, and plasmin. All the proteolytic enzymes decreased viscosity, and as expected, hydrolysed the protein. They had no measurable effect on deoxyribonucleic acid. The deoxyribonucleases (dornases) decreased viscosity only very slightly, had no proteolytic effect, but naturally depolymerised the deoxyribonucleic acid. Plasmin, lysozyme and hyaluronidase demonstrated no effect on any parameter measured. The proteolytic activity of trypsin was enhanced somewhat if combined with dornase.

Gessert then followed up these *in vitro* tests with systemic administration, first of chymotrypsin by buccal administration, to six patients, followed by an enteric coated trypsin-chymotrypsin tablet in a further ten patients. Buccal administration did not improve the clinical picture, but four out of ten patients given trypsin-chymotrypsin orally showed improvement of symptoms; however, in only one instance did the effusion disappear. Chemical analysis failed to indicate significantly increased hydrolysis of the protein.

Muftik (Muftik, 1957) treated a wide variety of pathologic secretions in cavities in the ear, nose and throat, with trypsin, chymotrypsin, papain, and streptokinase-streptodornase. Subjectively, a high proportion of patients' conditions were cured or ameliorated, and the author concluded that the treatment of chronic disease was often shortened and the indication for radical surgery rendered unnecessary in 75% of cases.

Chronic middle ear effusions are responsible for a high proportion of cases of hearing loss in children. A particularly common, and difficult to manage type of effusion is characterised by tenacious and thick exudate. This type is aptly named 'glue ears'. The direct instillation of chymotrypsin solution via the intact tympanic membrane using a fine hypodermic needle has been used to liquefy the viscous exudate (Litton and McCabe, 1962). In most cases, the liquefied material was spontaneously discharged through the Eustachian tube. Hearing was immediately improved and no adverse reactions were seen in the 31 patients taking part in the trial.

Coyas has confirmed this work and extended the indication to catarrhal otitis media which also proved successful in normalising hearing and reducing exudate after a course of antibiotics had failed (Coyas, 1963).

Later Mawson (Mawson,1967) introduced the α-chymotrypsin solution into the middle ear via a myringotomy incision; the liquefied exudate was forced down the Eustachian tube with a current of air induced by a Politzer's bag. A cure rate of 76% was found.

3.6.3 Sinusitis

Acute and chronic sinusitis are characterised by accumulation of excessive amounts of vicous mucus in the maxillary sinuses, and often in the ethmoid and frontal sinuses. The production of this thick tenacious material is a result of inflammation and results in the characteristic pain and stuffiness which accompany the condition. Both acute and chronic sinusitis are commonly accompanied by infection, and in addition to draining the sinuses to relieve the unpleasant symptoms, treatment with antibiotics may be required.

Chymotrypsin administered by injection, buccally, or mixed trypsin-chymotrypsin tablets was given to a total of 26 patients with chronic sinusitis in uncontrolled clinical trials (Billow *et al,* 1960; Teitel *et al,* 1960). Subjective complete relief of symptoms was shown in 10 patients, partial relief in 9, and no response in 7. In one of these trials, those patients who did not completely respond to one trypsin-chymotrypsin tablet four times a day were given twice the dose, when complete recovery was shown in 87% of cases, and the remainder partly recovered. Even better results may be obtained if the proteolytic enzymes are combined with antibiotic (Gordon *et al,* 1970). (See Section 3.9, 'Drug Potentiation Effects').

In a clinical study on 25 patients with chronic sinusitis who were given bromelain tablets, improvement in symptoms was shown in 64% of those treated (Hine, 1966).

A double blind clinical evaluation of bromelains was conducted in 48 patients with acute sinusitis or nasal mucosal inflammation, breathing difficulties and associated headache. Response to treatment was measured by subjective assessment of pain relief and easing of breathing problems; more objective scoring of reduction in nasal mucosal inflammation and discharge was also undertaken. Mucosal inflammation was reduced in 83% of those patients receiving bromelain compared to a 52% of those who were given placebo tablets. Breathing difficulty was relieved in 78% of those on bromelain treatment and in 68% of those on placebo. Overall 87% of patients on active bromelain tablets showed good to excellent results compared to 68% of those receiving placebo (Ryan, 1967).

3.7 FIBRINOLYSIS

Thrombotic diseases are responsible for an increasing morbidity and mortality in Western style cultures. The reduction or elimination of clots restricting circulation is an urgent necessity to reduce the possible damage to the patient or perhaps mortality arising from acute thrombo-embolic incidents. Treatment with anti-coagulants allows normal physiological processes to restore the circulation, but is inevitably slow, while surgical intervention is dramatic, and on occasions, a procedure involving a finite risk to the patient.

The possibility of lysing fibrinous emboli with proteolytic enzymes occurred to a number of early investigators who mainly used trypsin (Innerfield *et al,* 1952) but also, in animal experiments, chymotrypsin, carboxypeptidase, papain, ficin, plasmin and streptokinase-streptodornase (Ambrus *et al,* 1957). Trypsin *in vivo* was not particularly effective and proved dangerously toxic when administered intravenously. In artificially induced thrombi in dogs, only plasmin proved effective. These dog experiments were deceiving since although streptokinase is ineffective in activating plasminogen in the dog, it is highly effective as a human plasminogen activator.

Plasmin is the major natural fibrinolytic enzyme product in the blood, and as such is the obvious choice for thrombolytic therapy. Alcohol fractionation techniques developed by Cohn (Cohn *et al,* 1946), and subsequently refined, have made plasminogen available from donated human plasma. This pre-enzyme may be activated to plasmin for therapeutic use. Some initial use of such an enzyme in 40 patients was described by Cliffton (Cliffton, 1957). Twelve of these patients suffering from venous thrombosis responded well, as did three patients with pulmonary emboli. Arterial or cerebral thrombosis did not improve.

Plasmin is expensive; the source of raw materials, donated or purchased human plasma, is limited. Furthermore, it has been shown that activation of plasminogen to plasmin within the clot is more effective than bathing the surface with the activated enzyme (Sherry *et al,* 1959). The blood also contains high levels of anti-plasmins which rapidly neutralise the enzyme following intravenous administration. *In vivo* plasminogen activators, therefore would appear to offer promise of greater effectiveness than the administration of exogenous plasmin.

In 1944, Kaplan (Kaplan, 1944) and Christenson (Christenson, 1944) showed that extracts of haemolytic streptococci contained a kinase capable of activating the plasminogen of certain species, including humans, to plasmin. Numerous animal experiments followed, but in 1959 Johnson and McCarty proved that streptokinase was effective in lysing artificially produced thrombi in human volunteers (Johnson and McCarty, 1959).

Streptokinase activates plasmin via a plasminogen-kinase complex which in the presence of additional plasminogen forms plasmin. Early preparations of streptokinase contained tangible amounts of impurities in the form of streptococcal proteins. These impurities caused distressing side effects such as pain at the site of injection, and feverish reaction in at least ten per cent of cases. The enzyme has been extensively purified, and the commercial preparations now have a offered reduced tendency to pyrogenic reactions, but being a bacterial protein, remain antigenic. Pyrexia still presents a problem during prolonged therapy; the reason for this occurrence is not known.

Streptokinase is inexpensive, and in plentiful supply. The enzyme may be given by intravenous or intra-arterial infusion for pulmonary embolism, or the treatment of deep vein thrombosis. The normal dose is 100 000 to 600 000 units, which is continued in accordance with the patient's needs. The enzyme has a half-life in blood of less than 30 minutes. After administration, part of the dose may be blocked by circulating antibodies (Bachmann, 1968), therefore an adequate initial loading dose is needed to neutralise these inhibitory substances.

The above information will suffice to show that although effective, streptokinase is not the ultimate *in vivo* plasminogen activator.

Work by Williams (Williams, 1951) and Astrup and Sterndorff (Astrup and Sterndorff, 1952) demonstrated that the fibrinolytic activity of urine was due to the presence of a plasminogen activator rather than a direct proteolytic effect on fibrin. In 1963 first adminstration in man was reported on volunteers under carefully controlled conditions with experimentally induced thrombi. They obtained complete resolution of thrombi in 9 out of 13 cases within 25 hours (McCarty *et al,* 1963). This initial study has since been confirmed, in particular by Sherry (Sherry, 1970) and in the U.S. pulmonary embolism trials (Urokinase Pulmonary Embolism Trial, 1973; Urokinase-Streptokinase Pulmonary Embolism Trial, 1974).

Urokinase, although more expensive than streptokinase, has a number of established and theoretical advantages. The enzyme has a low toxicity, it does not cause allergic reactions or stimulate antibody formation, and it is normally non-pyrogenic. Unlike streptokinase which is non-selective and causes general plasminogen activation in the circulation, urokinase has a higher affinity for plasminogen in the gel phase within a thrombus. In theory, therefore, urokinase should be less likely to initiate bleeding complications.

In spite of the theoretical greater safety of urokinase, all enzyme thrombolytic therapy is contra-indicated in patients with healing wounds of any type - surgical, diagnostic or accidental. Both *in vitro* and *in vivo* observations suggest that urokinase or streptokinase therapy is more effective if initiated while the clot is fresh and before the fibrin structure has become organised.

Effectiveness of streptokinase or urokinase therapy in pulmonary embolism (Urokinase Pulmonary Embolism Trial, 1973; Urokinase-Streptokinase Pulmonary Embolism Trial, 1974; Miller *et al,* 1971; Frantantoni *et al,* 1975) or deep vein thrombosis (Kakkar *et al,* 1969; Common *et al,* 1976; Seaman *et al,* 1965) have been confirmed by numerous authors. It has not yet been definitely established whether enzyme thrombolytic therapy leads to lower mortality rates than conventional anti-coagulant treatment, and in all cases the particular benefit appears to lie in the resorption of large fresh thrombo-emboli that are life-threatening or are likely to cause serious prolonged vascular incompetence. When it is combined with heparin, mortality is reduced compared to heparin treatment alone (Brochier, 1978). Use of streptokinase to treat cardiac infarction has been less conclusive. A European Working Party reported a mortality rate of 19% in 357 patients treated with streptokinase within 24 hours compared to 27.4% mortality in 339 similar patients treated with heparin. Reinfarction was less common in the group treated with streptokinase (*Report of European Working Party,* 1971). Similar results were published by Bett (Bett *et al,* 1973) following a multicentre study when 227 control patients with cardiac infarcts were treated with warfarin and heparin and compared to 230 patients treated with warfarin, heparin and streptokinase.

Streptokinase therapy is more successful in retinal artery occlusion. Twelve out of twenty-five patients were improved in an uncontrolled study (Rossman, 1973) and in a controlled study 20 patients with central retinal vein occlusion treated with streptokinase followed by anti-coagulant therapy showed significantly better vision after 12 months than the control group. Unfortunately, permanent blindness due to vitreous haemorrhage occurred in three patients and this was attributed to the streptokinase therapy. This factor must limit its use in retinal artery occlusion (Kohner *et al,* 1976).

Successful treatment of central retinal vein thrombus in elderly patients with oral bromelains has been reported (Gray, 1969).

New approaches to thrombolytic therapy have been reviewed recently (Kakkar and Scully, 1978). These include intermittent streptokinase infusion to reduce the influence of circulating inhibitors, combination of streptokinase with the anti-coagulant heparin, which reduces formation of new fibrin on the existing clot, and combinations of streptokinase with exogenous plasminogen or plasmin.

3.8 ANTI-INFLAMMATORY EFFECTS

3.8.1.1 Animal and clinical pharmacology studies

The anti-inflammatory effect of trypsin and other proteolytic enzymes was observed by Martin *et al,* (Martin *et al,* 1954) following a series of investigations in animals which had received local administration of a number of phlogistic agents. The first approach was to observe the influence of various proteolytic enzymes on the oedema induced by egg white. The enzymes were trypsin, streptokinase, chymotrypsin, prolase B, and fibrinolysin; all were effective, and a dose response related activity was observed. Martin also compared the effect of fully activated trypsin with the completely inactivated enzymes and concluded that the anti-inflammatory action is a function of its proteolytic activity.

The inhibition of the local oedema produced by dextran injection is very dependent on dose and relative time of administration. High doses of trypsin given 24 hours before induction of oedema are most effective (Martin, 1957).

In rats, very low doses of trypsin proved effective in inhibiting oedema produced by yeast (Adamkiewicz *et al,* 1955).

Having established activity in purely local oedema, Cohen (Cohen *et al,* 1955) and later Martin (Martin, 1957) studied the effects of subcutaneously injected trypsin on generalised oedemas produced by the intraperitoneal injection of egg white and dextran. Higher doses of trypsin were required to control this type of oedema, compared to those needed to reduce purely local inflammation. It has also been proved in animals that the anti-inflammatory action of trypsin is independent of the adrenal cortex since it is retained after adrenalectomy and hypophysectomy (Adamkiewicz *et al,* 1955; Cohen *et al,* 1955).

Martin's initial studies have indicated that an anti-inflammatory action could be a general property of most proteolytic enzymes, comparable effectiveness being dependent on relative dosage. Continuing research has confirmed that in addition to trypsin most proteases show an inhibitory action on artificially induced oedema and inflammation in animals. Gordon and Ablondi (Gordon and Ablondi, 1957) induced inflammation in rabbit eyes, and showed a decreased severity of reaction when streptokinase was administered intravenously.

Miechowski and Ercoli (Miechowski and Ercoli, 1956) showed that the inflammatory response in rats, mice and guinea pigs to turpentine, anaphylactic challenge, ultraviolet radiation and thermal burns was inhibited better by chymotrypsin than trypsin. This was confirmed using egg white oedema in rats (Calnan *et al,* 1963) and in rhesus monkeys whose limbs had been damaged by crush injury. Animal studies have also confirmed that bromelains, papain and ficin exhibit similar anti-inflammatory properties.

In view of the success in animal experiments with trypsin and other proteolytic enzymes, Innerfield conducted a study in over 500 patients suffering from a wide variety of inflammatory and thromboembolic conditions. Trypsin was administered by slow intravenous infusion. Although this trial was primarily aimed at thrombolysis, considerable success in resolving inflammation was demonstrated (Innerfield *et al,* 1953). Although thrombolysis was achieved, this was subsequently realised not to be due directly to the trypsin administered, and Innerfield turned his attention to the anti-inflammatory action. Trypsin by intravenous injection often gave rise to uncomfortable and occasionally serious side effects, and in later clinical trials the enzyme was given by deep intramuscular injection as a suspension in oil or as a stabilised solution. Chymotrypsin had been demonstrated as effective in reducing experimental inflammation in animals and was known to be generally less toxic. It was a natural progression, therefore, to confirm its efficacy in man (Davis *et al,* 1959; Billow *et al,* 1960). These initial treatments used a suspension of the enzyme in oil; this was painful on injection, and was rapidly superseded by a lyophilised preparation for reconstitution with saline immediately before use. Chymotrypsin injections were effective, gave rise to less side effects than trypsin, but the risk of anaphylaxis in sensitive patients remained. Injections are also unpopular with many patients and, except in a hospital situation, inconvenient.

The ultimate aim for most drug formulators is to present their drugs in a convenient, compact oral dose form. Enzymes are no exception to this general aim, and as early as 1953, Innerfield was conducting clinical trials with buccal trypsin tablets (Innerfield *et al*, 1953), intended to dissolve slowly against the cheek for absorption through the buccal mucosa. In 127 patients, Innerfield showed a low incidence of side effects, and a rapid and consistent reversal of inflammation. No effect on plasma prothrombin, fibrinogen, antithrombin, coagulation time or fibrinolysis could be detected. It was concluded that buccally administered trypsin had no influence on clotting mechanisms.

Oral proteolytic enzyme therapy is an obvious progression, but some daunting theoretical obstacles present themselves. Acid pepsin is highly destructive to most proteinaceous enzymes and either digests them or induces irreversible inhibition. This can be overcome by the use of gastric juice resistant enteric coatings. An alternative is to give papain which is unaffected by gastric acid pH, or to protect the enzyme with skimmed milk (Pecile *et al*, 1967).

A further possible objection is the traditional view that molecules of the size of proteolytic enzymes cannot be absorbed intact by the small intestine. Clinical experience confirmed the effectiveness of the oral route of administration for most proteolytic enzymes (See 3.8.3), but it remained to prove absorption. Bastian *et al*, (Bastian *et al*, 1956) and later Avakian (Avakian, 1964) showed that circulatory esterase levels can be measured following the oral administration of trypsin and chymotrypsin. This work has been confirmed by Ambrus *et al*, (Ambrus *et al*, 1967) using mixed trypsin - chymotrypsin tablets in volunteers and assessing changes in blood esterase levels. Further evidence of absorption of orally administered proteolytic enzymes has been provided by radioactive tracer studies (Moriya *et al*, 1967; Martin *et al*, 1957; Kabacoff *et al*, 1963). In one study in rats (Moriya *et al*, 1967) the radioactive chymotrypsin fraction isolated from the blood was checked to confirm that proteolytic activity had been retained during the absorption process.

One final controversial point remains. Even if absorption can be shown to occur, the blood contains massive amounts of circulating inhibitors, sufficient to overwhelm the amounts of enzyme likely to be absorbed from an oral dose. These inhibitors are non-specific and will inhibit most therapeutically administered proteases *in vitro* (Tan and Gans, 1969). Although alpha-one anti-trypsin inhibits enzyme activity completely, alpha-two macroglobulin only reduces the activity of the enzymes by approximately 25% (Mehl *et al,* 1966) and the enzyme thus complexed is not further inhibited by other circulating inhibitors. Alpha-two macroglobulin may, therefore, act as a carrier for absorbed enzymes.

Interest in the effects of orally administered proteolytic enzymes on serum inhibitors and other plasma proteins has stimulated a number of investigations, partly in the hope of throwing light on the exact mode of action of these drugs on the inflammatory process (See 3.8.2).

The evidence which has emerged is somewhat conflicting and almost certainly requires further investigation and explanation. Injections of chymotrypsin into patients following mastectomy gave a further increase in already elevated chymotrypsin inhibitor levels, although oral administration of the enzyme did not (Silverman *et al,* 1966). Similarly intramuscular injection of trypsin into rats gave rise to a significant increase in trypsin inhibitor levels, the maximum being attained on the third day (Hladovec *et al,* 1957).

Margetts and co-workers determined changes in total trypsin-inhibiting capacity, and specific acute phase reactive protein levels in 46 surgical patients, approximately half of whom received active trypsin/chymotrypsin tablets; the remainder were either untreated or received dummy enzyme tablets on a double blind basis. It was concluded that there is a distinct early change towards normal, of one of the important acute phase reactants responsible for the inhibition of plasmin; alpha-one anti-trypsin. The reduction of this inhibitor was statistically significant on the fifth day post-operatively in the enzyme treated group. (Margetts *et al,* 1972).

Fisher *et al* conducted a similar series of estimations on surgical patients undergoing hernia repair. The methodology was slightly different, and this research group reported that although C-reactive protein, α_1-anti-trypsin, and total trypsin inhibitory capacity increased as anticipated in the control group of a double blind trial, the administration of pancreatic proteolytic enzyme tablets inhibited the rise of C-reactive protein but enhanced trypsin inhibitor capacity. (Fisher *et al,* 1974).

Many common anti-inflammatory drugs are protease inhibitors (Skidmore and Whitehouse, 1966) and, in fact, the pancreatic protease inhibitor, aprontinin, is claimed to have anti-inflammatory action, although the evidence is somewhat tenuous and not as well substantiated as that for the enzymes themselves.

Measurement of anti-inflammatory response in small animals is relatively simple, and the effectiveness of oral proteolytic enzymes in rats has been confirmed; in man, however, the question is more problematic. Acute traumatic inflammation is a condition that normally resolves itself fairly quickly, although considerable immobility, pain and general discomfort may be experienced for several days following the injury. Oedema in limbs may change its position and distribution, thus making measurement extremely difficult. It is logcal that a number of steps have been taken to try to substantiate that proteases are useful as anti-inflammatory agents in humans. Winsor (Winsor, 1972) suggested that differential skin temperatures between areas of skin burned with ultraviolet radiation and unburned areas might show an early difference between volunteers treated with proteolytic enzymes and a placebo treated group. A radiometer and a thermograph were used to measure temperatures. A significant difference in temperature differences was demonstrated between the enzyme treated subjects and those receiving placebo. This was highly significant for back burns at $P = <0.001$, and significant at $P = <0.01$ for abdominal burns.

A further development of this approach (Hambury *et al,* 1975) utilised a specifically designed electronic recording instrument so that the skin temperatures could be continuously monitored over a period. Differential skin temperatures were measured in two groups of patients suffering from Colles' fracture. One group received placebo tablets while the other was given trypsin-chymotrypsin tablets. In all patients a thermocouple junction was placed immediately over the fracture while another was placed on the forearm about 15 cm away. Differential temperature traces were taken over a period of 3-5 days. The enzyme treated group showed a tendency towards normal within 24 hours, whilst in the placebo group temperature fluctuation was still apparent after 3 days. As heat is one of the cardinal signs of inflammation, this more rapid normalisation of skin temperature with the enzyme treated group was taken as an objective indication of faster resolution of inflammatory oedema.

3.8.2 Mode of action

The exact mode of action of many drugs is elusive, and can defy explanation for years after therapeutic efficacy has been established beyond doubt. A particular problem presents itself in the case of the anti-inflammatory effects of enzymes because of the complexity of the inflammatory processes themselves, the exact biochemical and physiological pathways of which are still under active investigation.

The classical concept (Masson, 1972) of inflammation involves the release of vasoactive amines, for example histamine and serotonin, from damaged mast cells. These substances are responsible for the dilation of capillaries which characterise the onset of the inflammatory process. Enzymes, also released from the damaged cells, activate kinins which sustain capillary dilation and increase vascular permeability. This increase in vascular permeability leads to leakage of fluid with consequent oedema which is believed to be enhanced and sustained by fibrinous deposits in the capillary circulation, thus slowing return to the veins (Houck and Jacob, 1965). This simple outline of the inflammatory process is almost certainly a gross over-simplification, but together with some theories covering the reversal of the inflammatory process it is an essential foundation for attempting to understand the mode of action of proteolytic enzymes.

Control of the inflammatory process meeds a network of fibrin for the fibroblasts to commence their essential repair work; on the other hand persistence of excessive amounts of fibrin interferes with this process. Consequently, normal healing requires the intervention of plasmin as a fibrinolytic agent at a precise moment.

Severe inflammatory reactions are characterised by the increase of acute phase reactive proteins (Werner, 1969) in the plasma, some being anti-proteases including α_1-anti-trypsin, the body's main plasmin inhibitor. This increase in anti-proteases is possibly a compensatory mechanism stimulated by the release of proteolytic enzymes from the damaged tissue. One of the other acute phas reactants is C-reactive protein which may promote phagocytosis (Ganrot and Kindmark, 1969), a process which could be important in the healing process.

A direct effect of proteolytic enzymes on kinins or histamine has never been claimed as a mode of action. Hakim (Hakim *et al,* 1961 showed that chymotrypsin by intraperitoneal injection had no effect on histamine, although Winsor claimed to show reduction in kinin production following the administration of proteolytic enzymes (Winsor, 1972). Nor is the anti-flammatory effect mediated through cortisol release, since it has been shown in animals that the anti-inflammatory effect of trypsin is independent of the adrenal cortex, being retained after adrenalectomy and hypophysectomy (Adamkiewicz *et al*, 1955).

Although most proteolytic enzymes can be shown to depolymerise fibrin *in vitro,* a similar direct effect *in vivo* is unlikely. Innerfield was unable to demonstrate that trypsin had a direct fibrinolytic effect *in vivo* and in any event both trypsin and chymotrypsin and most other proteases used as anti-inflammatory enzymes are inhibited by circulating inhibitors in the blood that are usually raised above normal levels in patients suffering from trauma.

Innerfield (Innerfield, 1957) showed that in rabbits trypsin and streptokinase both augmented tissue depolymerase activity, but this was apparently not a direct effect. He accounted for anti-inflammatory activity by induced alteration in the shape of macromolecules, changes in tissue permeability and an inhibition of proteolysis preventing the deposition of fibrin. He observed that inflammation was resolved in peripheral areas even when thrombi were present in larger vessels. These observations tend to discount the direct fibrinolytic action theories proposed by Bastian and others (Bastian *et al,* 1956).

It has also been proved that bromelains are inactivated in serum both *in vitro* and *in vivo.* When 50 mg/kg of bromelain was given to human volunteers no increase in proteolytic activity could be detected, and very low levels were found after intra-duodenal administration of large doses of bromelains to cats. No direct or indirect activation of the fibrinolytic system could be detected, hence these exper iments tend to confirm that direct fibrinolysis is not the main mode of action.

As early as 1943 Grob (Grob, 1943) had reported on the effects of injections of trypsin in stimulating the production of anti-trypsin in animals. These observations have led to theories that the stimulation of a further increase in anti-proteases in the blood may reinforce the body's natural defence mechanism against the non-specific proteolytic enzymes produced by cell damage and which augment the inflammatory condition by various pathways. This possibility has recently been given additional credence by the work of Fisher *et al,* (Fisher *et al,* 1974), who showed greater increases in alpha-one anti-trypsin and total protease inhibitors in a group of surgical patients undergoing hernia repair who had been given mixed trypsin-chymotrypsin tablets, compared to a parellel group who received placebo tablets. In this trial the rise in C-reactive protein titres was inhibited by the enzyme treatment. Furthermore, rises in blood inhibitor levels to chymotrypsin have also been demonstrated as a result of an injection of the enzyme into patients undergoing radical mastectomy. Oral administration in this instance did not appear to stimulate additional inhibitory activity (Silverman *et al,* 1966).

A slightly different indirect effect associated with anti-protease levels has been proposed by Margetts *et al* (Margetts *et al,* 1972). In their double blind trial on surgical patients they showed an early decrease in α_1-anti-trypsin and total protease inhibitors following the administration of trypsin-chymotrypsin tablets. They proposed that this decrease in serum inhibitory activity reduced tissue anti-plasmin levels and so enabled this enzyme to act on fibrin blocking the micro-circulation. Thus oedema clears quicker and the healing process can commence at an earlier stage. One interesting aspect of this work is that the statistically significant change in inhibitor levels occurred between the third and fifth days following operative injury, which coincides well with a number of published clinical papers. It is not clear whether this reduction in inhibitors is due to binding of absorbed enzymes onto circulating inhibitory proteins, or an effect on liver synthesis.

A further possibility exists. The effects of common anti-inflammatory drugs on prostaglandin synthesis are well documented. Until a very short time ago, no work has been conducted on the action of proteolytic enzymes on prostaglandins and associated substances. It has now been shown in tissue culture experiments that when intact monolayers of human cultured endothelial cells are incubated with thrombin or trypsin, prostacyclin (PGI_2) is released. Undisturbed monolayers or those treated with A.D.P. or epinephrine do not produce significant amounts of PGI_2. PGI_2 is an unstable prostaglandin which inhibits platelet aggregation and serotonin release and will inhibit polymorphonuclear leucocyte function. The authors suggest that the ability of various proteolytic enzymes to induce PGI_2 synthesis may represent a modulating mechanism in the control of inflammatory processes (Weksler *et al,* 1978).

3.8.3 Clinical uses

In spite of the difficulty in defining exact modes of action, the clinical effectiveness of proteolytic enzymes in treating and in the prevention of a wide variety of inflammatory conditions has been proved beyond reasonable doubt.

Many proteolytic enzymes of differing substrate specificity have been shown to possess anti-inflammatory action. The main ones are trypsin, chymotrypsin, bromelains, papain, streptokinase, and ficin.

Netti, Bandi and Pecile compared the antiphlogistic profile of a number of these proteases with each other and with known anti-inflammatory drugs on a series of different types of rat paw oedema. Pancreatic proteolytic enzymes, bromelain and ficin demonstrated similar high anti-oedema activity when administered orally in a special gastroprotective vehicle in high dosage. This activity was similar for all types of oedema except that induced by brewers' yeast injection. Papain and bacterial protease proved largely ineffective. With the exception of aspirin, the synthetic drugs had a lower spectrum of activity (Netti *et al,* 1972).

There is, therefore, a variety of enzymes to be considered but also a great variety of clinical conditions. The evidence published is in fact so overwhelming that it is essential to be selective, and describe in detail only examples which demonstrate a particular aspect.

Proteolytic enzymes have been claimed to be effective in the treatment of soft tissue injuries, sciatica due to disc herniations and severe back pain, fractures, surgical procedures such as hand surgery, episiotomies, vasectomies, vein stripping and dental surgery.

The historical injections of trypsin as a suspension in oil (Innerfield, 1954), aqueous solutions of chymotrypsin (Fullgrabe, 1957; Nechtow and Reich, 1960; Teitel *et al,* 1960; Streiker, 1961) and streptokinase (Miller, 1957) have been largely replaced by oral administration, and most of the proteolytic enzymes claimed to be effective anti-inflammatory agents are available in tablet form.

The following examples of published clinical investigations serve to illustrate the type of evidence available to confirm the general effectiveness of the different proteases currently in use.

Matta and Mouzas (Matta and Mouzas, 1972) treated 60 patients suffering from a variety of haematomas, ecchymoses, sprains and contusions, but without fractures or lacerations, with streptokinase-streptodornase tablets for five days. Each tablet contained 10 000 units of streptokinase and 2500 units of streptodornase, and were given four times a day. A similar group of 60 patients, but not treated with enzyme tablets, were used as matched controls. Of the enzyme treated patients 82% improved in the first week compared to 28% in the control group. A statistically significant difference between the results of the two groups was shown.

Blonstein (Blonstein, 1969) investigated the effects of bromelain tablets on 74 boxers with bruising, and compared results with 72 controls. At the end of four days, all signs of bruising had cleared completely in 58 of the treated group compared to 10 in the control group.

In a double blind cross over study (Tassman *et al,* 1965), using a placebo control on 16 patients undergoing dental surgery, bromelain tablets were shown to be superior to placebo in 13 out of 16 cases The degree and duration of both swelling and pain were reduced.

Cirelli (Cirelli, 1967) reporting five years' clinical experience with bromelains in more than 700 patients suffering from oedema and inflammation of various origins, was impressed by the marked reduction in pain, oedema and inflammation and a decreased period of disability resulting in a quicker return to normal activities.

In a double blind study (Holt, 1969) on 125 patients with various sporting injuries, 65 were treated with papain tablets and 60 with an inactive placebo. Of the enzyme treated subjects 70% showed a better than expected therapeutic response compared to 20% in the placebo group. Although the methods of assessment were subjective the difference was statistically significant. Other double blind trials in rhinoplasty (Vallis and Lund, 1969) and head and neck surgery (Lund and Royer, 1969) confirmed the effectiveness of papain tablets compared to placebo.

By far the greatest volume of published evidence exists for the use of trypsin and chymotrypsin combination tablets in the treatment of oedema and inflammation of varying origins and the acceleration of healing of traumatic injury.

The following more recent double blind and controlled trials illustrate the effectiveness of pancreatic proteases given orally in controlling inflammation.

A double blind controlled trial (Buck and Phillips, 1970) was carried out on 61 professional footballers from two football clubs to assess the value of trypsin-chymotrypsin tablets in the resolution of inflammation and haematoma associated with injuries sustained while playing football. The injured players received either proteolytic enzyme tablets or an identical placebo, and all were also given the normal palliative measures and physiotherapy in addition to the tablets. The relative recovery times of the active and placebo groups were assessed and a significant difference was found between players who had recovered on days 3 and 4 after treatment commenced. It is interesting to note the coincidence in time of recovery of most players, with the significant reduction in plasma proteases inhibitors reported by Margetts *et al* (Margetts *et al,* 1972).

Fifty patients with injuries to muscles, tendons, and fractures were treated with trypsin-chymotrypsin tablets or a placebo on a double blind basis (Schwinger, 1970). The degree of swelling and ecchymosis was evaluated, using a grading system: 65% of enzyme treated cases showed a satisfactory response to therapy compared to only 4% in the placebo group. A rapid decrease in swelling and reduction in pain characterised the enzyme treated group.

Sillar and Mouzas (Sillar and Mouzas, 1973) showed that the deformity which commonly follows Colles' fracture may be minimised by the use of trypsin-chymotrypsin tablets, and in fractures of the hand the rate of recovery is increased and the degree of swelling and pain reduced (Shaw, 1969).

Surgical operations such as vein stripping are associated with a haematoma formation due to the considerable tissue trauma. A double blind trial on 100 patients undergoing operations for varicose veins demonstrated the effectiveness of trypsin-chymotrypsin tablets in reducing the haematoma. In this case, the difference between the active and placebo groups did not become significant until the thirteenth day (Rinisten, 1971).

Severe low back pain and sciatica are common conditions, treatment being limited to rest, manipulation, immobilisation, analgesics or skeletal muscle relaxants. Three fairly recent papers (Hingorani, 1968; Gaspardy *et al,* 1971; Gibson *et al,* 1975) have shown the effectiveness of trypsin-chymotrypsin tablets in reducing the unpleasant symptoms associated with severe back pain and intervertebral disc herniation. It is believed that the proteolytic enzymes reduce the inflammatory oedema at the spinal nerve root, thereby reducing nerve compression and decreasing the associated pain.

It is interesting to note that chymopapain by direct injection into the intervertebral region has been successful in dissolving and thus removing a displaced intervertebral cartilage and thereby relieving the symptoms of nerve compression (Sullivan, 1975).

The majority of trials, which now cover several thousand patients, have shown a demonstrable effectiveness for trypsin-chymotrypsin administered orally. As might be expected, a few trials have reported a lack of significant effect (Craig, 1975; Forrest *et al,* 1968). Both of the trials considered here were double blind and well conducted, one in sprained ankles and the other in dental surgery. The lack of demonstrable effect illustrates the difficulties associated with the objective measurement of clinical response in acute inflammatory conditions.

3.9 DRUG POTENTIATION EFFECTS

An interesting potential application for proteolytic enzymes, which would repay further detailed investigation, is the property of these substances to potentiate the activity of a number of drugs including antibiotics.

The effect was first observed and reported by Seneca and Peer in 1963 (Seneca and Peer, 1963). In a group of 26 patients a 40 mg dose of chymotrypsin in enteric coated tablets was given two hours before a 250 mg oral dose of tetracycline. Average increases in tetracycline blood levels of 55% were observed after two hours and 29% after four hours. In a second group of only four patients the two drugs were given simultaneously. In this case, the enhancement of blood levels was even greater, being by 233% at two hours after drug administration and 115% after four hours. The authors concluded that the increased blood levels were due to an increase in intestinal absorption.

The same investigators continued their research and gave a second report about a year later (Seneca and Peer, 1964). In these experiments, the chymotrypsin and tetracycline administration was continued six-hourly for twenty-four hours, and in a further group of patients the chymotrypsin was given for twenty-four hours at six-hour intervals followed by a single intramuscular injection of 100 mg of tetracycline. In all cases blood levels of tetracycline were increased substantially. In this second experiment, urinary excretion of tetracycline was measured and found to be increased by 244% in two hours and 131% in four hours. Intramuscular injection of tetracycline and chymotrypsin concomittantly into rabbits also showed significant increases in blood levels (Seneca and Peer, 1964).

As a final confirmation of their previous work, Seneca and Peer (Seneca and Peer, 1965) published a couble blind cross-over trial, in which the relative effects of 20 mg or 40 mg of chymotrypsin plus 250 mg of oral tetracycline were compared in the same patients who received dummy tablets plus 250 mg tetracycline. A further small group received active or placebo chymotrypsin and 100 mg of tetracycline by intramuscular injection. Serum tetracycline levels in all patients were measured at two-hourly intervals both microbiologically and chemically. Mean serum tetracycline levels were 45% higher in the subjects when they received 20 mg chymotrypsin plus tetracycline, than when they received the tetracycline plus placebo tablets. When the dose of chymotrypsin was increased to 40 mg a mean increase of 50% in serum concentration of the antibiotic was recorded, (Figs. 3.1 and 3.2). Urinary excretion of tetracycline was also increased by 40 and 52% at the two chymotrypsin dose levels.

When tetracycline was administered intramuscularly and 40 mg chymotrypsin was given orally, the mean urinary antibiotic levels increased by 192% compared to tetracycline alone.

Although the authors conclude that the chymotrypsin enhances the absorption of tetracycline from the intestinal tract, their observations with the antibiotic by intramuscular injection seem to indicate that this may be an over-simplification of the effects observed, since these observations have been contested by Scioli and Ciaffi (Scioli and Ciaffi, 1965). It is possible, for example, that intestinal absorption is not actually increased, but the higher blood levels and urinary excretion are due to displacement of protein bound antibiotic resulting in higher concentration of free and microbiologically active tetracycline in the serum.

An additional indication that mere increase in intestinal absorption is not the complete picture was given by Thevenot and others (Thevenot *et al,* 1966). A study was conducted to observe the concentrations of tetracycline contained in inflammatory granuloma tissue and exudate and in the liver and lung tissues of rats receiving a single dose of tetracycline, compared to similar animals who received α-chymotrypsin in addition to the antibiotic. Both the chymotrypsin and the tetracycline were administered by intraperitoneal injection.

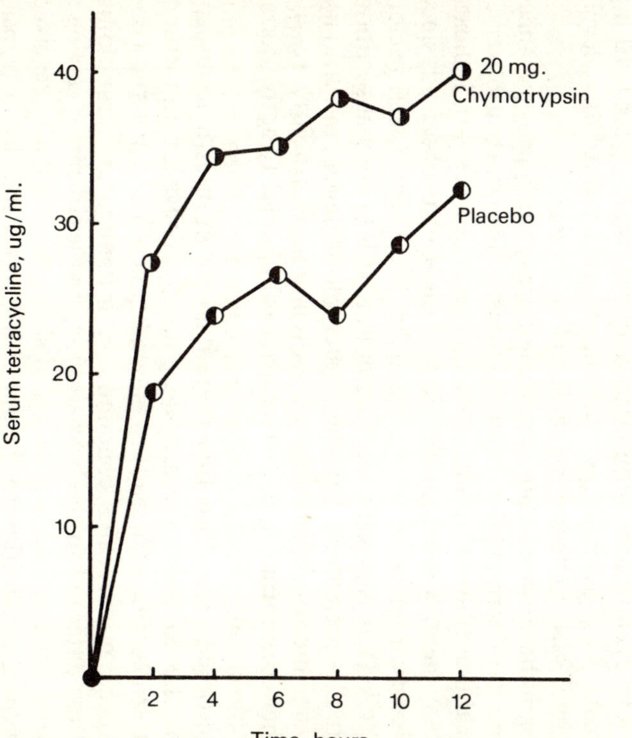

Figure 3.1

Serum Tetracycline levels of Group I subjects [*Reprinted by kind permission of the Journal of the American Geriatric Society Vol: 13; page 711, 1965*].

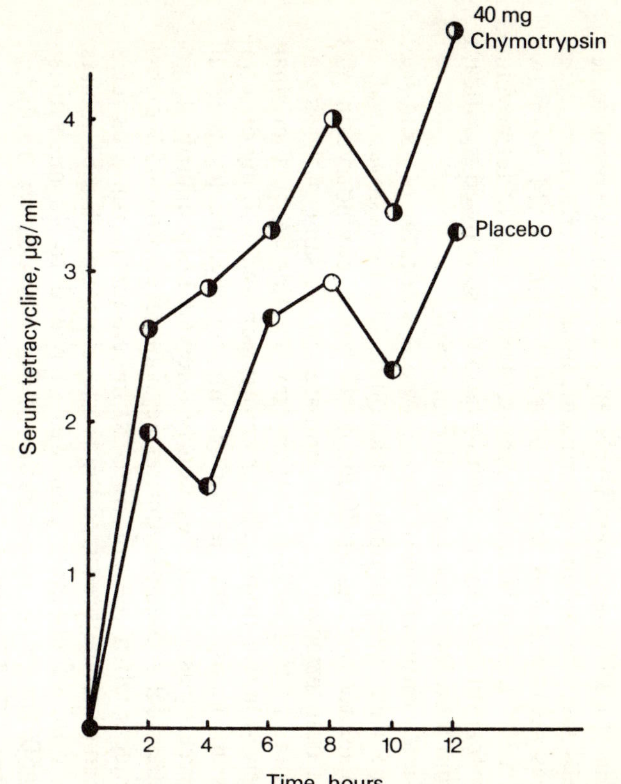

Figure 3.2

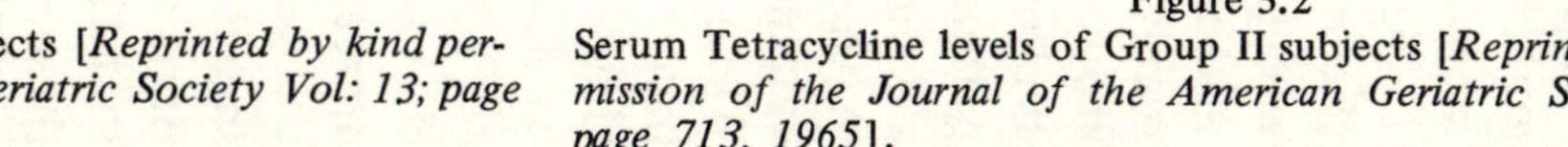
Serum Tetracycline levels of Group II subjects [*Reprinted by kind permission of the Journal of the American Geriatric Society Vol: 13; page 713, 1965*].

The granulomata were induced by the injection of croton oil. Because of possible difficulties in estimating tetracycline by microbiological or chemical methods, the antibiotic was tagged with tritium and the radioactivity in the tissues and exudates measured.

In animals receiving the chymotrypsin plus antibiotic, the levels of tetracycline in liver, lungs, granuloma and inflammatory exudate were higher than those receiving the antibiotic alone. It is suggested that either the chymotrypsin affects diffusion of the antibiotic from the injection site, or exerts a selective effect on the permeability of certain membranes thus allowing greater tissue penetration of the antibiotic. Other investigations have shown increased blood levels in rats, but not in granulomata (Pecile *et al,* 1967). The clinical significance of these biochemical and pharmacological observations has not been assessed.

One-hundred cases of pelvic inflammatory disease were treated in a double blind study (Bulwa, 1969) comparing tetracycline plus trypsin-chymotrypsin with tetracycline alone. The overall success rate with the enzyme plus antibiotic group was 92% compared with 68% successful treatment with antibiotic alone.

A further double blind trial in patients suffering from infected sinusitis was carried out by Gordon and co-workers (Gordon *et al,* 1970). The recovery rate was improved in the group of patients receiving tetracycline plus proteolytic enzymes (trypsin-chymotrypsin) compared to the similar group receiving tetracycline alone. The recovery rate was based on reduction of pain in the sinus area, reduction of volume of mucus and the change of mucus from purulent to mucoid. Although the effectiveness of the enzymes is attributed primarily to increased concentrations of antibiotic, their mucolytic effect is proposed as an additional advantage in the particular condition under treatment.

Tetracyclines have been shown to be effective in the treatment of acne vulgaris, partly it is believed, by their effect in influencing triglyceride hydrolysis and partly by reducing bacterial flora, particularly *Corynebacterium acnes.* The effectiveness of tetracycline is not uniform in all patients, owing to irregularity of absorption and penetration to the skin. Stankler (Stankler, 1976) treated 62 severe acne patients with either tetracycline alone, or tetracycline plus trypsin-chymotrypsin on a double blind basis. Of the patients on the combined treatment 69% were improved, compared to 52% taking the antibiotic alone.

In spite of the considerable experimental and clinical evidence supporting potentiation tetracycline antibiotic activity, some research teams have been unable to confirm the effect, or the exact mechanism of action in animals (Macdonald *et al,* 1965; Edwards, 1978). Edwards was unable to demonstrate increased absorption using an isolated intact gut loop in rats, nor an effect on drug clearance or protein binding (Edwards, 1978).

Some very recent further confirmatory clinical evidence has reconfirmed the increase in therapeutic efficacy of tetracycline plus trypsin-chymotrypsin over the same dose regime of tetracycline alone. Fifty patients with moderate to severe acne were treated in the trial. Half were given capsules containing 250 mg of tetracycline, the other half 250 mg tetracycline plus trypsin-chymotrypsin in an identical capsule. The treatments were coded and a double blind technique used in issuing the capsules. Photographs were taken of each patient initially and the severity of the acne was graded. At the end of a two-month period the acne was regraded for each patient and an assessment of change for better or worse made: 78.3% of patients were improved on the antibiotic treatment with enzymes compared to 43.5% on antibiotic alone (Liddell, 1978).

Increase in absorption or tissue penetration of other antibiotics have also been observed in animals. Wohlman, Syed and Ronchi (Wohlmann *et al*, 1968) demonstrated significantly higher serum, eye and brain levels of penicillin G in rats when the antibiotic was administered intramuscularly concomitantly with chymotrypsin by intraperitoneal injection. This confirmed earlier work by Avakian and Kabacoff (Avakian and Kabacoff, 1964) who combined oral administration of penethicillin and chymotrypsin and showed an increase in antibiotic blood levels.

Very significant increase in penicillin levels in granulomatous lesions and inflammatory fluids has also been achieved using intraperitoneal injection of chymotrypsin in rats prior to intramuscular injection of penicillin (Wohlman *et al,* 1969a,b). Control groups of animals received intramuscular penicillin alone. Granulomas were induced by the implantation of cotton pellets, and peritoneal inflammation was induced by injection of turpentine into the peritoneal cavity.

The authors state that preliminary investigations which they have carried out indicate that chymotrypsin increases penicillin penetration into macrophages cultured in Hanks medium, and that after addition of chymotrypsin, isolated kidneys and eyes become considerably more permeable to penicillin. These effects show that the effects observed in the rats can, at least in part, be attributed to increased cell permeability.

In vitro incubation of serum containing penicillin with chymotrypsin increases antibiotic activity. In this case, the chymotrypsin is probably displacing or releasing the penicillin from its protein bound state. Such a situation would further increase blood levels and tissue penetration of the antibiotic. Similar effects are observed when various antibiotics such as mono-mycin, tetracycline, erythromycin and levomycetin are incubated in a broth culture medium containing pathogenic *Staphylococcus aureus* and trypsin or chymotrypsin are added (Maloman *et al,* 1975; Belizhenko, 1975).

A Russian publication (Danyuschenkova *et al,* 1977) showed the superiority of lincomycin plus chymotrypsin over lincomycin alone or no treatment in three groups of albino mice infected with *Staphylococcus aureus.*

The enhancement of drugs by proteolytic enzymes is not confined to antibiotics. The synergistic anti-inflammatory activity of chymotrypsin and prednisolone has been described in rats and guinea pigs (Wohlman *et al,* 1969a,b). Inflammatory lesions were induced by a variety of agents including lithium chloride, lysozyme, methyl salicylate, histamine and dry ice. The animals were divided into untreated controls, and those who received (a) chymotrypsin intraperitoneally, (b) chymotrypsin orally, protected by reconstituted powdered milk, (c), (d) the two routes of enzyme administration, plus prednisolone alone.

Resolution of inflammation was recorded photographically. The most complete healing was achieved by treatment with combined chymotrypsin and prednisolone. This was significantly better than the healing resulting from the administration of either drug alone. It was concluded that in addition to the combined known anti-inflammatory effect of the two drugs, an increased local penetration of the steroid induced by changes in cell permeability due to the chymotrypsin could have been a relevant factor.

A 20% increase in blood levels of chlordiazepoxide in human volunteers has been demonstrated at the 10 mg chloridiazepoxide per day dose. Chymotrypsin had little effect at the 15 mg per day or 10 mg three times a day level of chlordiazepoxide, possibly owing to a saturation effect (Foldes *et al,* 1970).

Increased absorption of isoniazid was demonstrated by Wisniewski when the drug was given with proteolytic enzymes (Wisniewski, 1972).

The concomitant administration of cyclophosphamide with trypsin-chymotrypsin tablets to patients suffering from a variety of metastasing carcinoma is claimed to improve therapy. It has not been possible to demonstrate a significant increase in survival times, but quality of life has been improved in many patients thus treated, and the effect on secondary tumours has, on occasions, been dramatic

(Silver, 1977–78). It is possible that the enzyme increases either the blood levels of cytotoxic drug, or the tumour cell permeability to the compound, or affects protein binding.

In rats, enhancement of blood and tissue levels of cyclophosphamide have been shown, and the activation of synthesisokinins in tumours when the drug was administered concomitantly with trypsin (Buckzo, 1972; Buckzo and Wisniewski, 1975). A combination of these effects should markedly increase the effectiveness of the cytotoxic agent. This was later confirmed in rats with Guerin tumours (Buckzo and Popow, 1976).

Edwards and Calvert (Edwards and Calvert, 1977) have attempted to find an explanation for the apparent increase in therapeutic effectiveness of cyclophosphamide when given with trypsin-chymotrypsin tablets, using an intensive study of plasma levels of the alkylating agent and its metabolites in a small series of patients. These patients were treated with either cyclophosphamide alone or the cytotoxic drug plus the proteolytic enzyme. They were unable to show a significant influence on the disposition of cyclophosphamide. They concluded that the reported improvement in therapy must be due to some other effect.

Although animal evidence for the potentiation of the effects of various drugs with proteolytic enzymes is convincing, the exact mode of action remains indefinite and further investigation of the human situation would appear to be warranted. If the claims for increase in drug activity can be substantiated, lower, and hence less potentially toxic, doses of selected drugs could be given without loss of therapeutic action, or the therapeutic effectiveness of a given dose could be increased.

3.10 CANCER

In addition to the improvement of cyclophosphamide therapy with concomitant proteolytic enzyme administration mentioned under 3.9, proteolytic enzymes have found other uses in the clinical diagnosis and treatment of cancers.

Thornes *et al* reported that remission was achieved in 3 out of 5 patients with acute leukaemia when given brinolase and anti-neoplastic drugs. One patient on brinolase alone also showed remission (Thornes *et al* 1972). The effect of the enzyme appears to be via the induction of autocytotoxicity. The exact mechanism of this action is not clearly defined, and there are several possibilities. Brinase is known to interfere with cell membrane permeability to change cell surfaces and so expose agglutination sites so that the enzyme treated leukaemic cells should be more accessible to antibody. Brinase also lowers anti-plasmin levels, and plasmin is a direct cytotoxic agent; the cells destroyed release agents which may generate antibodies. Hence brinase may exert a double action of exposing antigens and enhancing antibody production.

Part of diagnostic procedures for detecting cancer of the stomach involves the recovery of cytologic material for differential staining by the Papainicolaou method and subsequent microscopy. Malignantcells can be readily identified. Various methods of freeing malignant cells have been used including brushes, sponges, and balloon catheters. Chymotrypsin solution introduced with a moderately forceful flushing action through a Levine tube, with subsequent aspiration, has been found to free cells which can be separated for microscopic examination by centrifugation (Rubin and Benditt, 1955).

The accuracy of this technique in diagnosis has been reported by Unicker *et al* as being 94.7% for positive diagnosis and 92.3% for negative diagnosis (Unicker *et al,* 1958). Sullivan indicates similar accuracy at 85% correctly diagnosed as malignant and 96% of cases correctly assessed as benign (Sullivan, 1960).

3.11 MISCELLANEOUS CLINICAL INDICATIONS

Proteolytic enzymes have been utilised for a number of clinical conditions outside those indications previously described. While many of these applications have remained largely experimental, several have enjoyed a period of established use until superseded by more effective drugs or allowed to lapse owing to changes in fashion of medical treatment.

Before the discovery of the current powerful vermifuge drugs, papain enjoyed a considerable vogue in combating worm infestation due to ascarid, oxyurid, and trichurid worms (Weise, 1950). The effectiveness of this enzyme is due to the fact that, while these parasites have powerful inhibitors to the normal proteases of the intestinal tract, trypsin and chymotrypsin, these inhibitors are ineffective against papain. The papain will, therefore, destroy the outer skin of the worms. Addition of the standard vermifuge, piperazine adipate, renders treatment even more effective. A single dose of papain is often sufficient, although the enzyme will not destroy the eggs.

Local injections of trypsin were suggested as first aid treatment for venomous snake bite, based on experiments with mice and dogs. When these animals had been infected subcutaneously with elapid venom followed by immediate local injection of trypsin all survived

(Hsiung *et al,* 1975). This possible treatment does not appear to have been confirmed in man.

Relief of dysmenorrhoea in a high proportion of cases has been found using the proteolytic enzyme bromelain. Local application of the enzyme induces dilation of the cervix and contraction of the uterine cornua (Hunter *et al,* 1960). The apparent safety of this enzyme when topically applied to the human cervix, and a spasmolytic effect, have been confirmed (Duncan *et al,* 1960).

Kallikreins, isolated from hog pancreas, have a marked hypotensive action. They liberate kinin peptides from kininogens which are present in the α-globulin fraction of the serum. These kinins increase capillary permeability, stimulate smooth muscle, and cause capillary vasodilation. The potential exists, therefore, for the use of kallikreins in peripheral vascular and coronary artery disease and migraine. The enzyme may also have value in increasing blood supply to fractures and injured tissue to promote healing.

Intravenous brinolase has proved effective in increasing the walking ability of 14 patients with intermittent claudication. A 100 mg dose of enzyme was administered weekly in sodium chloride injection initially, but the dose was later intensified by increasing daily injection.

3.12 FUTURE POSSIBILITIES

This chapter has shown the very extensive historical and present medical uses of proteolytic enzymes. As is the case with most drugs, importance of enzyme therapy in certain areas has declined as new drugs are discovered and improvements to therapy emerge. For example, pepsin as an aid to digestion is largely discounted, and the use of proteases in the direct liquefaction of empyema is a relatively rare procedure. On the other hand, the use of enzymes to remove vascular thrombi is of growing importance. Pancreatic proteolytic enzymes remain a vital adjunct to the treatment of cystic fibrosis.

In spite of the overwhelming evidence to support the clinical effectiveness of proteolytic enzymes in resolving inflammation and oedema, the mode of action is not yet established beyond doubt and would repay detailed investigation. Elucidation of the pathways by which enzymes exert this effect would further increase the understanding of the inflammatory process and could lead to more effective enzyme preparations and dose forms. Development of new proteolytic enzymes by fermentation process may also assist in achieving this objective.

Another potentially exciting area which has been little explored is the potentiation of drug penetration and clinical response when administered concomitantly with proteolytic enzymes. If this phenomenon is confirmed, it could lead to more effective therapy of antibiotics due to increased penetration to tissues of difficult access, or the greater pharmacological response at lower dosage to drugs which are toxic, with a resultant reduction in side effects.

Better formulation of pancreatic proteolytic enzymes, or the development of highly potent replacement enzymes by fermentation processes, could lead to improvement in the therapy of cystic fibrosis.

Pretreatment of proteins for dietary or therapeutic purposes is not new. The production of peptones for injection or incorporation into dietary supplements by digestion of lean beef with pepsin and pancreatin proteases is well established. A new development in this area has been suggested by Oliver *et al* who have purified dermal collagen grafts with trypsin and shown that the resultant material after cross linking with glutaraldehyde was not rejected. In rats, it was shown that the trypsin treated grafts had become recellularised and revascularised, and implants recovered up to 280 days after implantation showed little evidence of collagenolysis or cellular rejection (Oliver *et al,* 1977; Oliver *et al,* 1976; Oliver *et al,* 1972). The possible significance of this discovery in replacement materials for human tissue loss is considerable.

This chapter has dealt almost exclusively with the exogenous administration or application of proteolytic enzymes, although some enzymes given, for example streptokinase or urokinase, have activated enzyme systems in the body. It is possible that this approach, using the body's own endogenous enzymes in target areas, may fulfil an increasing role in future therapy. For example, in wounds and ulcers it is known that if the correct environment of precise temperature, moisture balance and mechanical cover, for example in the form of specially modified polyurethane film, can be provided, collagenases and proteases are activated. These will liquify necrotic tissues within and surrounding leg ulcers. The overall effect is cleansing, and a favourable situation for healing and epidermal overgrowth is provided (Lock, 1978).

Whether the future trend is towards improved proteolytic enzymes for direct administration, activation of enzymes within the patient, or merely the provision of the optimum environment or trigger to promote proteolysis, it seems certain that the therapeutic use of this important class of enzymes in medicine will continue for many years to come.

REFERENCES

Adamkiewicz, V. W., Rice, W. B., and McColl, J. D. (1955) Can. J. Biochem. Physiol. **33**; 332–339.

Allaben, R. D., Posch, J. L., and Larsen, R. D. (1962) *J. Bone Jt. Surg.* **44A**; 41–48.

Ambrus, J. L., Ambrus, C. M., Back, N., Sokal, J. E., and Collins, G. L. (1957) *Ann. N. Y. Acad. Sci.* **68**; 97–137.

Ambrus, J. L., Lassman, H. B., and de Marchi, J. J. (1967) *Clin. Pharmacol and Therapeutics.* **8**; 362–368.

Anson, M. L. (1939) *J. Gen. Physiol.* **22**; 79–89.

Astrup, T., and Sterndorff, I. (1952) *Proc. Soc. Exp. Biol. Med.* **81**; 675–678.

Avakian, S. (1964) *Clin. Pharmacol. Therap.* **5**; 712–715.

Avakian, S., and Kabacoff, B. L. (1964) *Clin. Pharmacol. Therap.* **5**; 716–720.

Bachmann, F. (1968) *J. Lab. Clin. Med.* **72**; 228–238.

Baker, L. (1951) *J. Biol. Chem.* **193**; 809–819.

Barraquer, J. (1958a) *Report to Roy. Acad. Med. Barcelona, April 8th.*

Barraquer, J. (1958b) Act. Ophth. (KBH) **36**; 803–806.

Barraquer, J. (1959) *Surv. Ophth.* **4**; 501–502.

Bastian, J. W., Hill, R. G., and Ercoli, N. (1956) *Proc. Soc. Exp. Biol. Med.* **92**; 800–803.

Belizhenko, G. G. (1975) *Antibiotiki* **20**:**9**; 823–825.

Bett, J. H. N., Biggs, J. C., Castaldi, P. A., Chesterman, C. N., Hale, G. S., Hirsh, J., Isbister, J. P., McDonald, I. G., McClean, K. H., Morgan, J. J., O'Sullivan, E. F., and Rosenbaum, M. (1973) *Lancet* (**i**) 57–60.

Billow, B. W., Cobodeville, A. M., Stern, A., Palm, A., Robinson, M., and Paley, S. S. (1960) *Southwest Med.* **41**; 286–287.

Billow, B. W., Stern, A., Oppenheim, A., Cabodeville, A. M., Martorella, F. S., and Paley, S. S. (1960) *Harlem Hosp. Bull. Feb. 41–45.*

Blonstein, J. L. (1969) *Practitioner* **203**; 206.

Boberg-Ans, J., and Badsberg, E. (1959) *Nord. Med.* **5**; 228–229.

Bourgois, P., and Bourgois, D. (1964) *Brit. J. Clin. Pract.* **18**; 533–534.

Bouvier, J. B., De La Barochez, H., Thenot, A., and Conil-Lacoste, F. (1962) *Therapie* **17**; 981–987.

Bruce, R. A., and Quinton, K. C. (1962) *Brit. Med. J. Feb. 3,* **Vol I**; 282–284.

Brochier, M. (1978) *Ann. Enesth. Franc.* **19(8)** 735–738.

Buck, J. E., and Phillips, N. (1970) *Brit. J. Clin. Pract.* **24**; 375–377.

Buckzo, W. (1972) *Ann. Med. Soc. Pol. Acad. Sci.* **96**; 285–287.

Buckzo, W., and Popow, J. (1976) *Acta. Physiol. Pol.* **27(4)**; 388–394.

Buckzo, W., and Wizniewski, K. (1975) *Acta. Physiol. Pol.* **26**; 429–432.

Bulwa, F. M. (1969) *Med. Dig.* **14**; 210–211.

Burke, J. F., and Golden, T. (1958) *Am. J. Surg.* **95**; 828–842.

Calnan, J., Kulatilake, A. E., and Saad, M. N. (1963) *Brit. J. Surg.* **225**; 743–750.

Camarata, S. J. (1956) *Dis. Chest* **29**; 388–399.

Cayle, T. (1963) *J. Soc. Cosm. Chem.* **14**; 249–259.

Christenson, L. R. (1944) *J. Bact.* **47**; 471–472.

Christie, R. B. (1960) *Pharm. J.* **II**; 508.

Christie, R. B. and Chudzikowski, R. J. (1971) *Manf. Chem.* **42**; 47–51.

Cirelli, M. G. (1967) *Clin. Med.* **74**; 55–59.

Claes, M. M. and Legrand, N. (1958) *Bull. Soc. Belg. Ophthal.* **119**; 491–498.

Clemmensen, I. and Christensen, F. (1976) *Biochim Biophys. Acta.* **429**; 591–599.

Cliffton, E. E. (1957) *Ann. N. Y. Acad. Sci.* **68**; 209–229.

Cogan, J. E. H., Symons, H. M., and Gibb, D. C. (1959) *Brit. J. Ophthal.* **43**; 193–199.

Cohen, H., Graff, M., and Kleinberg, W. (1955) *Proc. Soc. Exptl. Biol. Med.* **88**; 517–519.

Cohn, E. J., Strong, L. E., Hughes, W. L. Jr., Mulford, D. J., Ashworth, J. N., Melin, M., and Taylor, H. L. (1946) *J. Amer. Chem. Soc.* **68**; 459–475.

Common, H. H., Seaman, A. J., Rosch, J., Porter, J. M., and Dotter, C. T. (1976) *Angiology* **27**; 645–654.

Cornbleet, T. and Chesrow, E. J. (1960) *Arch. Dermatol.* **82**; 261–262.

Costa, P. N., and Fontana, P. F. (1959) *La Clinica Bologne* **19**; 321–323.

Coyas, A. (1963) *J. Laryngol. and Otology* **77**; 1001–1005.

Craig, R. P. (1975) *Injury* **6**; 313–316.

Cunningham, L. (1965) *Comprehensive Biochemistry* **16**; 85–188.

Danyuschenkova, W. M., Kardovich, G. A. and Berenshtein, T. F. (1977) *Antibiotiki* **23**; 330–333.

Davis, P., Levine, A. J., Beck, C., and Horowitz, B. (1959) *Post. Grad. Med.* **26**; 719–723.

Di Magno, E. P., Malagelada, J. R., Go, V. L. W., and Moertel, C. G. (1977) *New Eng. Med. J.* **296 (23)**; 1318–1322.

Dimick, A. R. (1977) *J. Trauma* **17**; 948–955.

Doleschel, W. (1975) Wien Klin. Wschr. **87**; 282–284.

Drummond, S., Saunders, J. H. B., Leach, R., and Wormsley, K. G. (1977) Scot. Med. J. **22**; 221–224.

Duncan, S. L. B., Lawrie, J. H., and MacLennan, H. R. (1960) *Lancet Dec. 31,* 1420–1422.

Ebata, M. and Yasunobu, N. (1962) *J. Biol. Chem.* **237**; 1086–1094.

Eck, R. V. and Dayhoff, M. O. (1966) *Atlas of Protein Sequence and Structure, National Biomedical Research Foundation, Silver Spring, Maryland p. 64.*

Edwards, I. G. (1978) *Ph.d. Thesis Manchester University.*

Edwards, I. G., and Calvert, R. T. (1977) *J. Pharm. Pharmacol.* **29** (Suppl.) 29P.

El-Gharbawi, M., and Whitaker, J. R. (1963) *Biochemistry* **2**; 476–481.

Fertck, O., Horehled, J., and Kolinsky, J. (1968) *Am. Perf. Cosm.* **83**; 33–34.

Firat, T. and Tinaztepe, B. (1970) *Acta. Ophthalmol.* **48**; 3–13.

Fisher, J. D., Weekes, R. L., Curry, W. M., Hrinda, M. E., andRosen, L. L. (1974) *J. Med.* **5**; 258–273.

Foldes, E. G., Campbell, K. N., and Wohlman, A. (1970) *Int. J. Clin. Pharmacol.* **3**(4); 335–343.

Forrest, W. I. N., Goodridge, D. L., MacDonald-Watson, A., and Starkey, W. E. (1968) *Brit. J. Oral Surg.* **6**; 7–10.

Francois, J., and Victoria-Trancosa, V. (1968) *Am. J. Ophthalmol.* **65**; 674–678.

Fratantoni, J. C. Ness, P., and Simon, T. L. (1975) *N. Eng. Med. J.* 1073–1078.

Fullgrabe, E. A. (1957) *Ann. N.Y. Acad. Sci.* **68**; 192–195.

Gambos, G. M. and Oliver, M. (1967) *Am. J. Ophthalmol.* **64**; 68–70.

Ganrot, P. O. and Kindmark, C. O. (1969) *Scand. J. Clin. Lab. Invest.* **24**; 215–219.

Gaspardy, G., Balint, G., Mitusova, M., and Lorincz, G. (1971) *Rheum. Phys. Med.* **11**; 14–19.

Gessert, C. F. Baumann, E. S. and Sentivia, B. H. (1960) *Ann. Otology. Rhinol, Laryngol.* **69**; 936–955.

Gibson, T., Dilke, T. F. W., and Grahame, R. (1975) *Rheumatol Rehab.* **14**; 186–190.

Gonda, I. (1977) *J. Pharm. Pharmacol.* **29**; 250–252.

Goodchild, M. C., Sagaro, E., Brown, G. A., Cruchley, P. M., Jukes, H. R., and Anderson, C. M. (1974) *Brit. Med. J.* **3**; 712–714.

Gordon, B. (1975) Brit. J. Clin. Pract. *29;* 143–146.

Gordon, S., and Ablondi, F. B. (1957) *Ann. N.Y. Acad. Sci.* **68**; 89–96.

Gordon, N., Thomas, J. V., Robinson, B. L., Wilson, R., McCarthy, J., and Bain, R. R. (1970) *Med. Dig.* **15**; 732–738.

Gray, L. F. (1969) *Sth Med. J. Nashville* **62**; 11–16.

Grob, D. (1943) *J. Gen. Physiol.* **26**; 405–442.

Habeeb, W., Reiser, H. G. Dick, F., and Roettig, L. C. (1954) *Dis. Chest* **26**; 408–419.

Hakim, A. A., Dailey, J. P., and Lesh, J. B. (1961) VII[e] Congres International De Therapeutique "Les Enzymes en Therapeutique" 106–107 Pub. Editions Medicine et Hygiene, Geneve.

Hambury, H. J., Watson, J., and Toole, A. (1975) *Med. Biol. Eng. March,* 202–208.

Hamit, H. F. and Upjohn, H. L. (1958) *Ann. Surg.* **147**; 580.

Hamit, H. F., and Upjohn, H. L. (1960) *Ann. Surg.* **151**; 589–593.

Harris, R., Norman, A. P., and Payne, W. W. (1955) *Arch. Dis. Childhood* **30**; 424–427.

Heimburger, N., Haupt, H., and Schwick, M., (1971) *Proc. Int. Conf. on Proteinase Inhibitors Eds. Fritz, H. and Tschesche, H., De Gruyter Berlin.*

Heinicke, R. M. and Gartner, W. A. (1957) *Econ. Botany* **11**; 225–234.

Hellgren, L., and Vincent (1977) *J. Int. Med. Res.* **5**; 334–337.

Hendley, C. D., Robins, K. C., Dertinger, B., Clements, G. R., and Ercoli, N. (1956) *Arch. Int. Pharmacodyn.* **106**; 164–166.

Hine, S. (1966) *Otolaryngol. (Tokyo)* **38**; 439–441.

Hingorani, K. (1968) *Brit. J. Clin. Pract.* **22**; 5–6.

Hladovec, J., Mansfield, V., and Horakova, Z. (1957) *Experientia* **14**; 146–147.

Holt, H. T. (1969) *Curr. Ther. Res.* **11**; 621–624.

Houck, J. C., and Jacob, R. A. (1965) *Exc. Med. Int. Conf.* **82**; 44–51.

Howes, E. L., Mandl, I., Zaffuto, S., and Ackermann, W. (1959) *Surg. Gyn. Obstet.* **109**; 177–188.

Hsiung Yü-Liang, Tsan Ju-Chin, Hou Yi-Ti, Liu Tz U-Chüan,Chou Hsing-Liang and Li Ching-Yen (1975) *Scientia Sinica* **18**; 396.

Hunter, R. G., Heinicke, R. M. and Civin, W. H. (1960) *Amer. J. Obstet. Gynec.* **79**; 428–431.

Innerfield, I., Schwarz, A., and Angrist, A. (1952) *J. Clin. Invest.* **31**; 1049–1055.

Innerfield, I., Angrist, A., and Schwarz, A. (1953) *J. Amer. Med. Assn.* **152**; 597–605.

Innerfield, I. (1954) *Surgery* **36**; 1090–1100.

Innerfield, I. (1957) *Ann. N.Y. Acad. Sci.* **68**; 167–175.

Irgang, S. (1960) *Harlem Hosp. Bull.* **1**; 19–24.

Jansen, E. F., and Balls, A. K. (1941) *J. Biol. Chem.* **137**; 459-460.
Jenkins, B. H. (1960) *Southern Med. J.* **53**; 44–46.
Jenkins, B. H. (1969) *Medical Times* **87**; 1613–1615.
Johnson, A. J., and McCarty, W. R. (1959) *J. Clin. Invest.* **38**; 1627-1643.
Kabacoff, B. L., Wohlman, A., Umhey, M., and Avakian, S. (1963) *Nature* **199**; 815.
Kakkar, V. V., Flanc, C., Howe, C. T., O'shea, M., and Flute, P. T., (1969) *Br. Med. J.* **1**; 806–810.
Kakkar, V. V., and Scully, M. F. (1978) *Brit. Med. Bull.* **34**(2); 191–199.
Kaplan, M. H. (1944) *Proc. Soc. Exptl. Biol. Med.* **57**; 40–44.
Karani, S., Kataria, M. S. and Barber, A. E. (1971) *Brit. J. Clin. Pract.* **25**; 375–377.
Kinnear-Wilson, A. B. and Stevenson, F. H. (1957) *Lancet Oct.* **26**; 820–823.
Knill-Jones, R. P., Pearce, H., Batten, J., and Williams, R. (1970) *Brit. Med. J.* **4**; 21–24.
Kohner, E. M., Pettit, J. E., Hamilton, A. M., Bulpitt, C. J., and Dollery, C. T. (1976) *Brit. Med. J.* **I**; 550–553.
Koke, M. P. (1967) *Am. J. Ophthal.* **63**; 1706–1709.
Konotey-Ahulu F. I. D. (1972) *Lancet* **II**; 714–715.
Krizek, T. J., Robson, M. C. and Groskin, M. G. (1974) *J. Surg. Res.* **17**; 219–227.
Lang, H., Helger, R., Lücker, P., Breddin, K., Hausamen T-U., and Rick, W. (1969) *Arzneimittel Forsch.* **19**; 939–944.
Lesurk, A., Terminielto, L., Travers, J. H. and Groff, J. L. (1967) *Fed. Proc.* **26**; 647.
Leuterer, W. (1957) *Medizinische* 310–314.
Levick, P. L., Brough, M. D., Vasilescu, C. T. and Laing, J. E. (1977) *Burns* **4**; 281–284.
Lewis, G. P. (1963) *Ann. N.Y. Acad. Sci.* **104**; 236–249.
Liddell, K. (1978) *Practitioner* **221**; 783–786.
Litton, W. B. and McCabe, B. P. (1962) *Laryngoscope* **72**; 182–187.
Lock, P. (1978) *Personal Comm.*
Lund, M. H. and Royer, R. R. (1969) *Arch. Surg.* **98**; 180–182.
MacDonald, H., Place, V., Diermier, H., Dornbush, A., Forres, M. and Kulkarni, S. (1965) *Antmicrob. Agents and Chemother.* **5**; 173–178.
Maloman, E. N., Syrbu, V. T. and Lupashku, B. K. (1975) *Antibiotiki* **20**:7; 613–617.
Mantegazza, A., and Rovesti, P. (1958) *Les Nouvelles Esthétiques* **6**;

12–15.
Margetts, G., Barber. K., Christie, R. B., Jones, W. E., and Bowden, W. T. (1972) *Brit. J. Clin. Pract.* **26**; 293–298.
Martin, G. J., Brendel, R. and Beiler, J. M. (1954) *Proc. Soc. Exptl. Biol. Med.* **86**; 636–638.
Martin, G. J. (1957) *Ann. N.Y. Acad. Sci.* **68**; 70–88.
Martin, G. J., Bognor, R. L., and Edelman, A. (1957) *Amer. J. Pharm.* **129**; 386–392.
Massan, P. (1972) *Louvain Med.* **91**; 61–78.
Matta, M., and Mouzas, G. L. (1972) *Practitioner* **209**; 343–345.
Mawson, S. R. (1967) *J. Laryngol. Otot.* **81**; 147–150.
McCarty, W. R., Newmann and Mecker (1963) in *"Pharmaceutical Enzymes"* Pub. Story Scientia (1978) p. 135.
McConn, J. D., Tsuru, D., and Yasunobu, K. T. (1964) *J. Biol. Chem.* **239**; 3706–3715.
Mehl, J. W., Park, M. Y., and O'Connell, W. (1966) *J. Soc. Exp. Biol. Med.* **122**; 203–210.
Mendel, L. B., and Blood, A. F. (1910) *J. Biol. Chem.* **8**; 177–213.
Miechowski, W. L. and Ercoli, N. (1956) *J. Pharmacol. Exptl. Ther.* **116**; 43–44.
Miller, J. M., Pierce, E. C. and White, B. H. (1953) *Milit. Surg.* **113**; 270–277.
Miller, J. M. (1957) *Ann. N.Y. Acad. Sci.* **68**; 185–190.
Miller, J. M., Kirkpatrick, H., Wooten, J. L., and Long, P. H. (1953) *N.Y. Med. J.* **53**; 2789–2795
Miller, J. M., White, B. H. and Long, P. H. (1953) *Postgrad. Med.* **13**; 438–442.
Miller, G. A. H., Sutton, G. C., Kerr, I. H., Gibson, R. V. and Honey, M. (1971) *Brit. Med. J.* **2**; 681–684.
Mohnke, W. (1953) *Med. Klin.* **48**; 1937–1938.
Morani, A. D. (1953) *Plast. Reconstr. Surg.* **11**; 372–379.
Moriya, H. Maniwaki, C., Akimoto, S., Yamaguchi, K., and Iwadare, M. (1967) *Chem. Pharm. Bull.* **15**; 1662–1664.
Moss, J. N., Frazier, C. V. and Martin, G. J. (1963) *Arch. Pharmacodyn* **145**; 166–189.
Muftic, M. K. (1957) *Ind. J. Med. Sci.* **11**; 1015–1020.
Murachi, T., and Neurath, H. (1960) *J. Biol. Chem.* **235(1)**; 99–110.
Nava, C. (1969) *Medna. Lav.* **60**; 732–734.
Nechtow, M. J., and Reich, W. J. (1960) *Amer. Practit.* **11**; 45–49.
Netti, C., Bandi, G. L., and Pecile, A. (1972) **Il**; *Farmaco* **27**; 453–466.
Neurath, H. and Schwert, G. W. (1950) *Chem. Rev.* **46**; 69–153.
Northrop, J. H. (1929–30) *J. Gen. Physiol.* **13**; 739–766.
Northrop, J. H., and Kunitz, M. (1932) *J. Gen. Physiol.* **18**; 433–458.

Northrop, J. H. (1931) *J. Gen. Physiol.* **15**; 29–43.
Ogston, D., Bennett, B., Herbert, R. J. and Douglas, A. S. (1973) *Clin. Sci.* **44**; 73–79.
Oliver, R, F., Grant, R. A., and Kent, C. M. (1972) *Br. J. Exp. Path.* **53**; 540–549.
Oliver, R. F., Grant, R. A., Cox, R. W. and Hulme, M. J. (1976) *Clin. Orthoped.* **115**; 291–302.
Oliver, R. F., Grant, R. A., Hulme, M. J. and Mudie, A. (1977) *Brit. J. Plast. Surg.* **30**; 88–95.
Parsons, D. J. (1958) *Clin. Med.* **5**; 1491–1494.
Pecile, A., Tosi, G. C., Bandi, G. L. and Previtera, M. A., (1967) *II Farmaco* **22**; 811–816.
Pecile, A., Tosi, G., Ferrario, G., Veronelli, C., (1967) *Il Farmaco* **22**; 583–589.
Peck, M. E., and Levin, S. (1952) *J. Thorac. Surg.* **24**; 619–636.
Pennisi, V. R., Abril, F., and Capozzi, A. (1975) *Burns* **2**; 169–172.
Prince, H. E., Etter, R. L., and Jackson, R. H. (1954) *Ann. Allergy* **12**; 25–29.
Profitos, J. (1965) *Med. Klin.* **60**; 548–550.
Prueter, R. D., Stein, M., Drum, J. A., and Kalz, F. (1957) *Canad. M. A. J.* **76**; 1040–1043.
Prytz, B., Connell, J. F., and Rousselot, L. M. (1966) *Clin. Pharm. Ther.* **7**; 347–350.
Ray, P. K. (1964) *Brit. J. Ophthal.* **48**; 230–231.
Read, C. T., and Berry, F. B. (1950) *J. Thorac. Surg.* **20**; 384–392.
Regan, P. T., Malagelada, J. R., and Go, V. L. W. (1976) *Clin. Res.* **24**; 567A.
Reiser, H. G., Patton, R., and Curtis, G. M. (1951) A. M. A. *Arch. Surg.* **63**; 568–575.
Reiser, H. G., Roettig, L. G., and Curtis, G. M. (1951) *"Forum of Fundamental Surgical Problems"*, 1950 Clinical Cong. of the American College of Surgeons, Philadelphia Pub. W. B. Saunders and Co. pp. 17–24.
Reiser, H. G., Roettig, L. G., Curtis, G. M. (1953) *Am. J. Surg.* **85**; 376–381.
Report of European Working Party (1971) *Br. Med. J.* (**3**); 325–331.
Rinisten, A. (1971) *Lakartidningen* **68**; 2381–2387.
Rodeheaver, G., Edgerton, M. T., Elliott, M. B., Kurtz, L. D., and Edlich, R. F. (1974) *Am. J. Surg.* **127**; 564–572.
Roettig, L. G., Reiser, H. G., Habeeb, W., and Mark, L. (1952) *Dis. Chest* **21**; 245–249.
Rossman, H. (1973) *Postgrad. Med.* **49**; Suppl. **5**; 105–108.
Rubin, C. E., and Benditt, E. P. (1955) *Cancer* **8**; 1137–1141.

Ruggiero, N. (1967) *Gazz. Med. Ital.* **126**; 1–12.
Ruyssen, R. and Lauwers, A. (Eds) (1978) *"Pharmaceutical Enzymes"* Pub. Story Scienta.
Ryan, R. E. (1967) *Headache* **7**; 13–17.
Saloman, A., Herchfuss, J. A., and Segal, M. S. (1954) *Ann. Allergy* **12**; 71–79.
Saunders, J. H. B., Drummond, S., and Wormsley, K. G. (1977) *Brit. Med. J.* **1**; 418–419.
Schwinger, O. (1970) *Wiener Md. Woch.* **36**; 1–6.
Scioli, C., and Ciaffi, G. (1965) *Antibiotica* **3**; 250–255.
Seaman, A. J., Common, H. H., Rosch, J., Dotter, C. T., Porter, J. M., Lawler, W. L., and Schlueter W. J., *Angiology* **27**; 549–556.
Seiter, H. H. (1955) *Dis. Chest* **27**; 179–189.
Seneca, H., and Peer, P. (1963) in *"Antimicrobial Agents and Chemotherapy; Ann Arbor Mich. Am. Soc. for Microbial."* **196**; 657–661.
Seneca, H., and Peer, P. (1964) *Scientific Exhibit. Amer. Urological Ass. Meeting, Pittsburgh, Pa. May 1964.*
Seneca, H., and Peer, P. (1965) *J. Am. Ger Soc.* **13**; 708–717.
Shaw, P. C. (1969) *Brit. J. Clin. Pract.* **23**; 25–26.
Sherry, S., Fletcher, A. P., and Alkjaersig, N. (1959) *Physiol. Rev.* **39**; 343–382.
Sherry, S. (1970) *J. Am. Med. Assn.* **214**; 2163–2172.
Sillar, R. W. and Mouzas, G. L. (1973) *Internat. Surg.* **58**; 31–33.
Silver, P. (1977–8) Personal Comm.
Silverman, I., Livingston, S. F. and Lipshitz, H. (1966) *Exp. Med. Surg.* **24**; 126–133.
Silverstein, P., Ruzicka, F. J., Helmkamp, G. M., Lincoln, R. A. and Mason, A. D. (1973) *Surgery* **73**; 15–22.
Simmonet, H., Thevenot, R., and Auclair, M. (1961) VIIe Congrès Internationale de Thèrapeutique, *"Les Enzymes en Thèrapeutique"* Pub. Editions Medicine et Hygiène Geneve 108–111.
Simon, S. W. and Harman, G. A. (1961) *J. Allergy* **32**; 493–500.
Skelton, G. S. (1968) *J. Chromatog.* **35**; 283–286.
Skidmore, I. F., and Whitehouse, M. W. (1966) *J. Pharm. Pharmacol.* **18**; 558–560.
Spencer, M. C. (1967) Cutis 3; 995–999.
Spier, I. R., and Clifton, E. E. (1954) *Surg. Gyn. Obst.* **98**; 667–674.
Spittler, A. W., and Parmenter, R. E. (1954) *J. Int. Coll. Surg.* **21**; 72–75.
Stankler, L. (1976) *Brit. J. Clin. Pract.* **30**; 65–66.
Startup, F. G. (1960) *Vet. Record* **72**; 245–246.
Startup, F. G. (1967) *J. Small Anim. Pract.* 689–691.

Steigrin, A. J. and Scott, C. S. (1952) J. A. M. A. **150**; 1403–1404.
Steinhardt, J. (1939) *J. Biol. Chem.* **129**; 135–144.
Streiker, F. B. (1961) *J. Am. Pod. Assn.* **51**; 713-718.
Stucke, K. (1949) *Chirurg.* **20**; 588–595.
Stucke, K. (1954) *Chirurg.* **25**; 289–294.
Sullivan, J. J. (1960) *J. Coll. Radiol. Aust.* **4**; 29–31.
Sullivan, M. (1975) *Proc. Roy. Soc. Med.* **68**; 479–480.
Tan, B. H. and Gans, H. (1969) *Pharm. Weekblad.* **104**; 717–721.
Tassman, G. C., Zafran, J. N. and Zayon, G. M. (1965) *J. Dent. Med.* **20**; 51–54.
Taub, S. J. (1960) *Clin. Med.* **7**; 2572–2577.
Teitel, L. H., Siegel, S. J., Tendler, J., Reiser, P., and Harris, S. B. (1960) *Indust. Med. Surg.* **29**; 150–152.
Thevenot, R., Roubertou, J., Uzan, A., Laine-Böszomenyi, M., Fallot, P., and Deiss, S. (1966) *Therapie* **21**; 457–472.
Thornes, R. D., Deasy, P. F., Carroll, R., Reen, D. J. and MacDonell, J. D. (1972) *Cancer Research* **32**; 280–282.
Tillett, W. S., and Sherry, S. (1949) *J. Clin. Invest.* **28**; 173–190.
Tillett, W. S., and Sherry, S. (1951) *J. Thorac. Surg.* **21**; 275–297.
Travis, J. and Liener, I. E. (1965) *J. Biol. Chem.* **240**; 1962–1969.
Trendtel, F. (1950) *Arztl. Wschr.* **5**; 659–662.
Troll, W., and Sherry, S. (1955) *J. Biol. Chem.* **213**; 881–891.
Unicker, W. O., Bolt, R. J., Hoekzena, A. D. and Pollard, H. M. (1958) *Gastroenterology* **34**; 859–866.
United States National Formulary (1975) *XIV*, 751–752.
United States Pharmacopeia (1975) *XIX*, 94.
Urokinase Pulmonary Embolism Trial (1973) *Circulation (Suppl)* **47**; II.
Urokinase-Streptokinase Pulmonary Embolism Trial (1974) *J. A. M. A.* **229**; 1606–1613.
Vallis, C. P. and Lund, M. H. (1969) *Curr. Ther. Res.* **11**; 356–361.
Weise, H. (1950) *Med. Clin.* **45**; 1096–1098.
Weksler, B. B., Ley, C. W., and Jaff, E. A. (1978) *J. Clin. Invest.* **62**; 923–930.
Werner, M. (1969) *Clin. Chim. Acta.* **25**; 299–305.
White, W. F., Barlow, G. H. and Mozen, M. M. (1966) *Biochem.* **5**; 2160–2169.
Wilde, N. J. and Derry, G. (1953) *Plast. Reconstr. Surg.* **12**; 131–137.
Williams, J. R. B. (1951) *Brit. J. Exp. Path.* **32**; 530–537.
Winsor, T. (1972) *J. Clin. Pharmacol.* **12**; 325–330.
Wisniewski, A. (1972) *Acta. Physiol. Pol.* **23**; 149–155.
Wohlman, A., Syed, M., and Ronchi, M. (1968) *Can. J. Physiol. Pharmacol.* **46**; 815–818.
Wohlman, A., Ramirez, R. I. and Avakian, S. (1969a) *Can. J. Physiol.*

Pharmacol. **47**; 301–304.
Wohlman, A., Syed, M., and Avakian, S. (1969b) *Experientia* **259**; 953–954.
Wright, L. T., Smith, D. H., Rothman, M., Metzger, W. I. and Quash, E. T. (1954) *J. Int. Coll. Surg.* **15**; 286–298.
Zawacki, B. E. (1975) *Surgery* **77**; 132–136.

Chapter 4

Solid Substrate Fermentation

Dr J. S. Knapp† and Dr J. A. Howell
Department of Chemical Engineering University College of Swansea.
Singleton Park, Swansea.

4.1 INTRODUCTION

The major difference between microbial growth on a soluble substrate and that on a solid substrate, is that with a completely dissolved substrate all of the substrate is equally accessible, whilst with a solid substrate much of it is initially inaccessible. During the fermentation the net amount of accessible substrate will always decline for dissolved substrates, but may either decline, increase, or remain constant at different stages of growth with solid substrates.

The factors affecting growth of micro-organisms on either insoluble substrates (both solid and liquid) or soluble materials which are polymeric and are incapable of permeating into cells, are less well understood than those affecting growth on soluble substrates. Workers dealing with these substrates often merely state that the organisms make use of extracellular enzymes. The enzymic degradation of insoluble materials is also not well understood, yet these materials are common in nature and form a large amount of the organic matter which is degraded every year in nature. Recently there has been increased interest in many microbiological processes involving insoluble or 'non-permeating' substrates. The production of SCP from hydrocarbons and cellulosic substrates, manufacture of commercial enzymes like proteases, amylases and cellulases, and the microbial leaching of mineral ores are examples of such processes which are of potential economic importance.

It is not the intention of this paper to go deeply into any of these particular fields, but rather to attempt to postulate the principles and fundamental similarities of microbial growth on solid substrates and to review experiments and kinetic models which describe such growth. The paper assembles information on a number of solid substrate fermentation systems, although the information available is far

† Current address: Dept. of Agricultural Biology University of Newcastle.

from equal in all the fields, and its nature depends on the discipline of the researchers. The literature concerning growth on hydrocarbon substrates is, for example, mainly quantitative and concerned with kinetics, whereas the literature on solid-state (Koji) fermentations and on wood rotting is mainly qualitative and contains almost no kinetic information.

In addition to reviewing the general features of growth on insoluble substrates this paper also considers in detail the fermentation of cellulosic substrates to produce single-cell protein.

4.2 GROWTH ON NON-PERMEATING SUBSTRATES

Many micro-organisms obtain their carbon and/or energy requirements for growth from materials which are insoluble or, if of low solubility, have such large molecules that they cannot freely pass through the permeability barrier of the cytoplasmic membranes (non-permeating substrates). Such substrates include proteins, certain carbohydrates, lignins, sulphur and some minerals. Work on the degradation of some of these substrates has been well reviewed, but this is not the case for all of them. Fermentation on hydrocarbon substrates is a special case where although the molecules can permeate the cell wall and membrane they are very sparingly soluble in an aqueous medium.

Solid and polymeric materials must be modified before they enter the cell and serve as carbon and energy sources. This modification is often a depolymerisation, but may be some other chemical process, as in microbial leaching. If this modification is enzymic it is reasonable to assume that the enzymes must act outside the cell. Pollock (1962), in a classic review, considered production of extracellular enzymes in bacteria. Many of his conclusions are still valid and can be applied equally to fungi. In particular, he laid down a classification of exo-enzymes dividing them into 1) Cell-bound enzymes which could be (a) intracellular or (b) surface-bound, and 2) Cell-free or truly extracellular enzymes. Pollock also dealt with the possibilities of autolysis (which would release intracellular enzymes) and the detachment of surface-bound enzymes, and suggested criteria to establish whether enzymes are truly extracellular, that is, liberated during normal active growth rather than as the result of cell damage or autolysis. With some substrates, such as cellulose, it can be very difficult to determine the location of exo-enzymes during active growth - and many authors do not attempt to. For some solid substrates, for example sulphur and metal sulphide ores, the agents which bring about the extracellular modification into a utilisable form may not be enzymes (Touvinen and Kelly, 1974).

The utilisation of solid substrates by microbes is affected by many varied factors, both physical and chemical, which are relatively unimportant to microbial growth on simple soluble substrates which can permeate the cell membrane.

4.2.1 Kinetics

It is possible to divide the kinetic problems of solid substrate fermentations into two groups - the first being concerned with the macromorphologyof the solid; its shape; surface to mass ratio; crystalline or amorphous nature; diffusivity within the solid; porosity; and with mass transfer to and from the solid surface. The above factors govern the rate at which whole molecules of substrate become accessible to enzyme attack. The second group of problems is associated with polymeric molecules having to be partially hydrolysed outside the cell into subunits which can pass through the cell membrane.

In all fermentations considered in this review the substrate must be modified outside the cell. This modification need not, however, require hydrolysis if the individual molecules can permeate the cell. It is thus worth examining in some detail fermentations in which only one of the two groups of problems mentioned is important. During fermentation hydrocarbons are generally thought to be broken down inside the cell (Goma *et al,* 1973) and thus show no group II effects, but they exhibit a number of the features of group I fermentation. The droplets are homogeneous and non-porous, and it is thus possible to study the effect of surface-to-volume ratio and mass transfer problems in isolation in respect to these fermentations. Fermentation of soluble cellulose such as CMC is a simple example of a pure group II problem. Microbial leaching is a group I problem if the solubilisation of the ions is considered to be due to changes induced in pH and other ion concentrations, that is by inorganic factors rather than enzymes. On the other hand if dissolution requires enzyme action the systems combine group I and group II effects. Group II effects are then those due to the kinetics of hydrolysis by the exocellular enzymes, the uptake by the cell of the permeable hydrolysis products and growth on these products.

Group I effects are those concerned with attachment of the cells to the substrate/medium interface to minimise transport delays and loss of enzyme or substrate; the diffusion of enzymes into the mass of solid the heterogeneity of the solid surface chemically and physically; and the accessibility of the solid to the cells themselves.

Clearly a comprehensive model would be pointless, but a number of successful models have been suggested of situations where only a small number of the above effects are quantitatively important.

The major factors, then, which influence the rate of a solid substrate fermentation are of a diverse nature and often have not all been quantified together in any model. The factors are summarised briefly below and are discussed in greater detail later.

4.2.2a Organism

The type of organism is important since even amongst organisms that ferment a given substrate the rates will differ greatly depending particularly on the mix of enzymes produced. In general with solid substrates the enzymes will be mainly non-constitutive, and thus catabolite repression and indusction will be important aspects of the overall kinetics. With a broad mix of enzymes it may be that repressive catabolites may not build up. On the other hand it may be advisable to introduce a second organism in order to metabolise the repressive catabolites of the first.

4.2.2b Enzyme kinetics

The kinetic rate of breakdown of the solid locally by the enzymes is a major factor influenced by substrate chemistry and enzyme mix. Product inhibition can effect the kinetic breakdown as can degree of polymerisation of the substrate. The solubility of the products of breakdown then governs the rate at which they can diffuse from the point of ezyme attack to the surface of the organism.

4.2.3 Substrate

4.2.3a Chemical factors

The chemical nature of the substrate is of major importance. If it is polymeric the degree of polymerisation and crystallinity are significant. The surface electrochemical properties are significant and thus whether it is hydrophobic and whether organisms will readily adsorb on to the surface.

4.2.3b Physical factors

(i) ACCESSIBILITY

The substrates physical morphology, especially porosity and particle size, will govern the accessible surface area to both organism and enzyme.

(ii) ORGANISM PROXIMITY

If the organism is close to the point of attack, as when it is adsorbed to the solid substrate, and especially if the enzymes remain attached to its surface, then the transport path for the breakdown products remains short. The concentration of the products at the cell surface is higher and growth is then faster unless modified by repressive effects of the products. Transport of the enzymes through the solid to the point of attachment is of importance when the cell is too large to penetrate there and wherever there is cell free enzyme present.

(iii) FILM EFFECTS

In some cases the organisms will grow as a layer or film over the solid surface to such a depth that the layer itself sets up a diffusion barrier to oxygen or nutrient penetration, and may result in local changes in pH causing local mineralisation which may enclose the organisms and bring a halt to the fermentation. Such an effect has been noted in microbial leaching.

(iv) MASS EFFECTS

In a solid state fermentation such as composting, the mass of solids acts as a barrier to oxygen transfer, moisture penetration, and also to heat. It is important that calculations are made in such fermentations to ensure that heat effects are controlled and that centre temperatures do not get too high and oxygen and moisture levels do not get too low.

4.3 GROWTH ON HYDROCARBONS

The production of SCP from hydrocarbons and the problems of marine oil pollution have led to much research on the growth of microbes on oil droplets (Einsele and Fiechter, 1975a; Erickson and Nakahara, 1975a). The hydrocarbons studied have ranged up to gas oil and waxes, but all have relatively small molecules compared to polysaccharides or proteins.

4.3.1 Mechanisms

Three possible substrate uptake mechanisms for microbes growing on a hydrocarbon in aqueous suspension have been identified although their relative importance is disputed. The first mechanism postulates direct contact through attachment of yeasts and bacteria to large oil droplets (Erikson *et al,* 1970; Miura *et al,* 1971a and b; Gutierrez and Erickson, 1977; Erickson and Nakahara, 1975; Mimura *et al*, 1971; Blanch and Einsele, 1973) or from attachment of small sub-micron 'accommodation' droplets to the cells (Einsele *et al*, 1975b; Aiba *et al*, 1969; Yoshida and Yamane, 1971b, 1974; Yoshida *et al*, 1973).

The second mechanism postulates uptake of dissolved oil, since there is a small but finite solubility (Yoshida *et al*, 1971a; Wodzinski and Bertolini, 1972; Wodzinski and Coyle, 1974) and the third mechanism postulates the transport of oil to the cell in small micelles of surfactants (Velankar *et al*, 1975). The second mechanism is often invoked for modelling uptake of sub-micron droplets called 'accommodation' droplets because they can be directly accommodated by the cell. These droplets then increase the apparent or pseudo-solubility of the oil.

4.3.1a Direct contact

In the direct contact mechanism the relative importance of sub-micron drops is uncertain. Gutierrez and Erickson (1977) observed that in an airlift fermenter $\frac{2}{3}$ of *C. lipolytica* cells were adsorbed to drops of n-hexadecane while the other $\frac{1}{3}$ were suspended in the aqueous phase, and were probably in contact with small 'accommodation' droplets. Yoshida *et al* (1973, 1974) considered that *C. tropicalis* obtained virtually all its substrate (n-hexadecane) from sub-micron droplets. It has been shown repeatedly that in a Stirred Tank Reactor (STR) the growth rate of microbes on liquid hydrocarbons in aqueous suspension is related to impeller speed and diameter (i.e. power input to liquid) (Moo-Young *et al,* 1971a and b; Wang and Ochoa, 1972; Blanch and Einsele, 1973; Aiba *et al*, 1969) and thus either to droplet size and therefore specific hydrocarbon interfacial area or possibly to floc formation with associated mass transfer limitations at low speeds (Blanch and Einsele, 1973). Specific hydrocarbon interfacial area has been found to be directly related to impeller speed, hydrocarbon concentration and surfactant concentration; and linear growth (Wang and Ochoa, 1972) will occur when either the total available interfacial area has been covered with cells, or when a mass transfer limitation has been reached.

4.3.1b Soluble hydrocarbons

It has been suggested that during growth of *C. tropicalis* on n-alkanes, substrate uptake due to direct contact with drops is negligible, the organisms presumably using dissolved hydrocarbon (Yoshida *et al*, 1971a). However, in a later report Yoshida and Yamane (1971b) showed that in their system (a biodisc with gaseous hydrocarbon feed) n-alkanes existed as sub-micron droplets – but they still considered direct contact to be unimportant in substrate transport. Wodzinski and co-workers (1972, 1974) consider that bacteria growing on several solid aromatic hydrocarbons in aqueous suspension take up substrate only from solution. The generation time is the same on dissolved hydrocarbon as in the presence of solid substrate, while the growth rate which is independent of the amount of solid present seems to be related to the solubility of the different substrates.

Organisms able to grow on hydrocarbons have been shown to have a greater chemical affinity for them than other organisms (Mimura *et al,* 1971; Miura *et al,* 1977a,b; Kaepelli and Fiechter, 1976); also glucose-grown cells of *C. tropicalis* adsorbed 25% less n-alkane than n-alkane-grown cells. Saturation of the cellular surface with substrate occurred in 30 seconds by a mechanism which was shown to be adsorption and not an enzymic reaction; it was thought that a lipopolysaccharide moiety present at the cell surface was responsible for cellular affinity to alkanes (Kaepelli and Fiechter, 1976), n-hexadecane-grown cells have twice the cellular lipid content of glucose-grown cells (Hug *et al*, 1974).

4.3.1c Pseudo-solubilisation

It has been observed that in hydrocarbon fermentations there is a change in the apparent substrate solubility during the course of growth, first increasing and later decreasing (Goma *et al*, 1973; Erickson and Nakahara, 1975a). This phenomenon known as pseudo-solubilisation may be due to surfactants which many workers have isolated, usually as lipoproteins or lipopolysachharides, from culture fluids of hydrocarbon fermentations (Zajic *et al,* 1977; Rapp and Wagner, 1976; Suzuki *et al*, 1969; Nakahara *et al*, 1977; and others, see Gutierrez and Erickson, 1977). Several authors have shown that artificial surfactants can increase the rate of hydrocarbon utilisation (Wang and Ochoa, 1972; Whitworth *et al*, 1973; Mimura *et al*, 1971).

Kinetic models have been produced to describe growth on liquid (Chakravarty *et al*, 1975) and solid (Chakravarty, *et al*, 1972) hydrocarbons, on the assumption that they are taken up from solution and that solution of the substrate is aided by a metabolite produced by the growing cells.

4.3.1d Micelles

Sufactants would, as well as forming micelles containing oil and increasing oil solubility, decrease the interfacial tension and thereby decrease the size of oil droplets (Gutierrez and Erikson, 1977) – increasing the total surface area. Agitation and surfactants therefore have similar effects. The increase in droplet surface area that they cause will increase the chances of cell collision with, and attachment to, droplets and also increase the rate at which hydrocarbons can pass into true solution. It seems likely that hydrocarbons reach the cell surface by all three mechanisms although their importance in quantitative terms is not easy to assess. The evidence for the different mechanisms is reviewed by Nakahara *et al*, 1977. Sometimes metabolisable hydrocarbons are dissolved in non-metabolisable materials; this is the case with gas oils. Under these circumstances diffusion of the substrate to the surface of the droplets is of importance (Erickson, *et al*, 1969; Erickson and Humphrey, 1969(b); Erickson *et al* 1970; Prokop *et al*, 1971). Where a pure substrate is in suspension it has been assumed (Erickson and Humphrey, 1969a) that the substrate can be represented during growth as a shrinking sphere with decreasing surface area. Stamatoudis and Tavlarides (1973) took into account the size distribution of the droplets and considered continued dispersal and coalascence. Their model when compared with experimental data showed that the presence of cells on the droplets sharply decreased coalescence efficiency. The work of Gutierrez and Erickson (1977) has shown that for *C. lipolytica* growing on n-hexadecane the interfacial area actually increases (2X) during the fermentation, only decreasing in the later stages. This increase is possibly brought about by surfactants and possibly by decreased droplet coalescence.

4.3.2 Mathematical methods

Several mathematical models have been produced which analyse the effects of different factors (Erickson,*et al*, 1969 a, b, c, 1970); Prokop *et al*, 1971; Moo-Young *et al*, 1971 a, b; Chakravarty *et al*, 1972, 1975; Aiba *et al* (1969); Miura *et al*, 1977).

Mallee and Blanch (1977) review the models that have been used and classify them into two groups. Group A deals with substrate uptake from sub-micron droplets in which the ultimate rate-limiting step is the diffusion of the droplets or of soluble hydrocarbon to the micro-organisms surface. In such models the amount of hydrocarbon existing in sub-micron droplets is termed its pseudo-solubility. Group B models deal with uptake from larger droplets by micro-organisms attached to their surface and in which the rate-limiting step is the available surface area on the droplets. As in any fermentation other soluble nutrients such as nitrogen can be made limiting, and in such cases the kinetics will be controlled by these nutrient concentrations and soluble substrate kinetics will apply. The accommodation droplets appear to be taken directly into the cell via lipid inclusions (Kennedy *et al,* 1975. Osumi *et al,* 1975) whence they are hydrolysed to fatty acids and then enter normal metabolic pathways. Apart from having a relatively high resistance to mass transfer the accommodation droplets once formed seem to behave as though they were a soluble substrate. The conversion of larger droplets to smaller droplets or solubilising process is aided by operating characteristics such as the agitation mode in the fermenter but is also affected by the organisms.

4.3.2a Complete model

Complete hydrocarbon models describe the fermentation by taking account of growth as a function of local chemical environment and distinguishing this from physical mechanisms responsible for transport. Empirical lumped parameter models account for growth as a function of the operating conditions.

In the continuous phase

Mass balance on substrate

Rate of increase of substrate concentration (1)	= − Rate of consumption by dispersed organisms (2)	+ Rate of transfer of substrate from droplets to continuous phase (3)

Mass balance on biomass

Rate of increase of mass of dispersed organisms (4)	= Rate of growth on continuous phase substrate (5)	+ Rate of detachment of organisms from droplets (6)
	− Rate of attachment of organisms to droplets (7)	

In or on the dispersed phase *Droplets in a given size class*

Mass balance on attached biomass

Rate of increase of organisms attached to class A droplets (8)	= Growth of organisms attached to class A droplets (9)	+ Rate of attachment of dispersed organisms to droplets (10)
	− Rate of detachment of organisms from droplets (11)	+ Transfer of cells to class A from other classes of droplets (12)
	− Transfer of cells from class A to other classes of droplet (13)	

Mass balance on substrate in droplets

Rate of increase of substrate in droplets of class A	= – Substrate consumption by organisms attached to droplet	– Rate of transfer to continuous phase
(14)	(15)	(16)
	+ Rate of transfer from other classes	– Rate of transfer to to other classes
	(17)	(18)

4.3.2b Discussion of terms

The terms on the left-hand sides (1), (4), (8) and (14) are simply the partial time derivative.

The local consumption terms are dependent on the growth rate of the micro-organisms and the concentration of substrate accessible to them. Terms (2) and (5) may include the concentration of pseudo-soluble substrate in the sub-micron 'accomodation' droplets. Terms (9) and (15) include the concentration of substrate in the droplet which may be pure substrate or a mixture of hydrocarbons. These terms also include a limit on the number of cells which may be in contact with the droplet, reached when its interfacial area is saturated with organisms. Generally Monod kinetics are chosen for the dependence of growth on local concentration.

The remaining terms are a function of agitation parameters, surface tension, impeller speed and for (3) and (16) (equal) the solubility of the hydrocarbon. Many physical assumptions have been made to create the different expressions of terms (6), (7) and especially (12), (13) (17) and (18). Simple models neglect the last four terms and assume a constant drop size distribution, but even that is influenced by agitation parameters and affects the interfacial area which appears in terms (9) and (15).

The details of the various models for droplet size distribution are perhaps unimportant here although it is significant that there is strong disagreement as to the form the equations should take. Stamatoudis and Tavlarides (1973) showed that the droplet size distribution could be skewed to lower droplet sizes than would be predicted in the absence of cells if it was simply assumed that coalescence efficiency was much less than one for droplets whose surface was covered with cells. In fact they assumed that only approximately five per cent of all droplet collisions led to coalescence. The observed increase in interfacial area during fermentation is thus explained by decreasing coalescence efficiency. Their predictions agreed with the experimental measurements of Wang and Ochoa (1972) but did not predict the formation of accommodation droplets. Terms (3) and (16) are generally assumed to depend on the product of mass transfer coefficient, interfacial area and concentration driving force, but Mallee and Blanch (1977) used a coalescence model to describe substrate transfer to microbial flocs. Chakravarty *et al* (1975) assumed that interfacial area was influenced *inter alia* by a substance released by the cells which changed surface tension. Actual models use simplifications of the full set of equations presented, and selected data can be matched with the simplified models.

The heterogeneous system has several features not found in the degradation of totally soluble substrates, and the second two mass balances are typical of those devised for insoluble substrates. The limited interfacial area accessible to the attached cells and the influence of cells on the transport processes are also general features.

Table 4.1

Terms in a typical model

Terms	Representation (in order)
1, 4, 8, 14	$\frac{\partial s}{\partial t}, \frac{\partial x}{\partial t}, \frac{\partial(\phi_i\alpha_i)}{\partial t}, \frac{\partial\psi_i}{\partial t}$
2, 5	$\frac{\mu_m S}{K_m+S} X$
9, 15	$\frac{\mu_m \psi_i}{K_m+\psi_i} \phi_i$
3 &16	$K_L \sum_i \alpha_i (\psi_i-S)$, $K_L \alpha_i (\psi_i-S)$
6 & 11	$K' \sum_i \phi_i \alpha_i$, $K' \phi_i$
7 & 10	$K' \sum_i (1-\phi_i)\alpha_i$, $K' (1-\phi_i)\alpha_i$
12, 13, 17, 18	$K_C \Sigma f(\phi_i, \phi_j, d_i, d_j)$

α_i interfacial area of droplet diameter d_i, in size class i.

ϕ_i fraction of surface covered by organism

ψ_i free substrate concentration in equilibrium with droplet size i

S free substrate concentration

X free cell concentration

4.4 GROWTH ON POLYMERIC SUBSTRATES

Much of the organic material found in nature is polymeric in structure, the four principal groups of polymers being polysaccharides, proteins, lignins and nucleic acids. Of these the first three are important and common as substrates for micro-organisms in nature. Cellulose, a universal structural material in plants, is the most abundant polymer. Other very abundant ones include pectates, hemicelluloses, alginates in sea weeds, chitin, lignins and certain structural proteins in animals such as collagen and keratin. The cytoplasmic membrane normally does not permit entry of large polymer molecules which are broken down external to the cell into diffusible subunits.

Large polymeric molecules are however produced by micro-organisms and transported from the inside to the outside of the cytoplasmic membrane. Evidence suggests that proteins probably cross the membrane in unfolded (linear) or partly folded form, only attaining their full tertiary structure after passing through the membrane (Costerton *et al,* 1974; Lampen, 1978).

4.4.1a Cell bound and cell free enzymes

Micro-organisms (excluding protozoa) degrading polymers use either enzymes released into the environment or enzymes bound in some way to the external surface of the cell. Attached enzymes can be released under circumstances such as cell death or environmental changes, resulting in a mixture of cell-free and cell-bound enzymes. Cell-free enzymes may strongly adsorb on an insoluble substrate and thus be undetectable in the culture filtrate. For a cell-bound enzyme to act on an insoluble substrate the cells must be in very close proximity to the substrate. The failure to find exo-enzyme activity in a culture filtrate may be due to the enzyme being cell-bound or substrate bound. It is also possible that an essential part of an enzyme (e.g. cellulase) may be cell-bound while the rest is cell-free. Gram positive bacteria tend to release a large portion of their degradative enzymes into the growth medium, whereas Gram negative bacteria (especially of wild type) tend to hold these enzymes in close association with the cell wall (Costerton *et al,* 1974; Rogers, 1970; Pollock, 1962; Lampen, 1965). This difference may be related to the basic differences in cell envelope structure between Gram positive and negactive bacteria, but Costerton *et al* (1974) point out its possible ecological significance. Gram negative organisms are well adapted to life in dilute aqueous environments, their wall-associated exo-enzymes do not become highly diluted, and the degradation products are produced in close association with the cells transport mechanisms thus reducing the risk of their loss by dilution in the medium. This may be the reason that the majority of bacteria found in river and sea water are Gram negative (Hodgkiss and Shewan, 1968). The Gram positive pattern of growth, with release of exo-enzymes, is more suited to growth in environments with more concentrated substrates.

4.4.1b Topological models

Hofsten (1975) has reviewed what he calls 'topological effects' in enzymic and microbial degradation. This refers to the spatial relationship between an organism, its substrate, and the degradative enzymes utilised. In order to develop complete kinetic models of the growth of microbes on polymeric substrates (like the models of Okazaki *et al* (1978 a,b) and Lee *et al* (1978) in which the concentration of cellulolytic enzymes is taken into account) it is necessary to have a clear understanding of these topological effects.

4.4.2 Substrate influence, cellulosic substrates

Physiochemical factors constraining the enzymic hydrolysis of, and thus microbial growth on, cellulosic substrates have been intensively studied, and a host of substrate pretreatments have been devised to overcome them.

4.4.2a Crystallinity

Cowling (1975) and Rowland (1975) have reviewed the factors which restrict cellulolysis. The chemical structure of the cellulose molecule is of prime importance; cellulose is a linear homopolymer of β–1,4 linked glucose molecules and is of indeterminate size; individual molecules are often well over 10^6 Daltons molecular weight. Many molecules aggregate into fibrils which have a very high degree of structural order; separate molecules are linked together by H-bonds to give a crystalline structure. In some parts of the fibrils the cellulose molecules are less linked and ordered. These non-crystalline regions are referred to as amorphous regions. It is generally held that the degree of crystallinity is the main factor in determining cellulose degradability, and the exact crystal structure (four forms are known) may be important (Cowling, 1975). Although highly crystalline cellulose is much more difficult to degrade than amorphous cellulose (Cowling, 1975; Dunlap *et al,* 1976) crystallinity may not be the causal factor. Stone *et al,* (1969) showed that crystallinity was not as important as accessibility, and a model of degradation of polymeric substrates (Okazaki and Moo-Young 1978 showed that the degree of polymerisation of the molecule (which is correlated with crystallinity) can account for many of the observed effects. Even altering the nature of the cellulase complex modelled by using different ratios of endo- and exco-cellulases caused changes in the degradation rates of high DP molecules similar to those observed when changing from an incomplete' to a 'complete' cellulase system.

4.4.2b Physical effects

(i) ACCESSIBLE AREA

As cellulose is insoluble it would be expected that its rate of hydrolysis would be dependent on the surface (rather than weight) available. This is certainly true but within limits, as has been pointed out by Stone *et al* (1969). Any new surface area generated by grinding or cutting etc. must be accessible to enzyme molecules. Methods which determine the area of surface available to gases are of no value in assessing area available to enzymes, and Stone *et al* (1968, 1969 a and b) and also Tarkow and Feist (1969) have developed methods to measure the accessibility of the substrate to macromolecules of similar sizes to enzymes. Stone *et al* (1969a) have clearly shown that the rates of cellulolysis are linearly related to the area of pore surface (or pore volume) available to macromolecules, which was correlated with the fibre saturation point of the substrate (cotton, swollen to varying degrees with phosphoric acid). Using spruce wood pulped to varying degrees Stone *et al* (1969a) showed that the rate of cellulolysis, fibre saturation point and median pore width all increased with increasing pulping yield (i.e. with decreasing lignin content). It was also shown that for all these parameters pulp that had been dried was inferior to pulp that had not been dried. It was concluded that drying did not affect the degradability of the accessible cellulose, but that by decreasing the mean pore diameter drying decreased accessibility and thus overall degradability. Similarly Mandels *et al* (1974) have found that after several different pretreatments fibres that had been dried were more resistant to cellulolysis than those that had not.

(ii) SUSCEPTIBILITY

With natural substrates which have not been pretreated it cannot be assumed that all the accessible area of substrate is equally susceptible to attack. This may be due to both physical and chemical factors such as crystallinity and lignin content. All the wood-rotting organisms attack wood cells from the lumen, to which they gain access by bore holes and enlarged pit canals (none seem to attack from the outside). This is presumably because the inner layers of the cell wall are more susceptible to enzymic attack than the outer layers. The S^2 layer seems to be more susceptible than the S_1 and S_3 layers, as shown by its preferential removal by soft, brown and sometimes simultaneous rot fungi.

(iii) SURFACE AREA

The most significant increase in area is probably that due to the exposure of the cell lumen. Owing to the greater susceptibility of the inner layers of cells it could be that the increased rate of cellulose hydrolysis due to this exposure will be disproportionately greater than the increase in surface area would suggest. Stone *et al* (1969a) suggest that repeated crossways cutting of woody cells would probably not greatly alter digestibility as, owing to small cell wall thickness compared to the cell length, only a relatively small increase in area would result. This assumption may not be valid, for several reasons. Enzymic attack is likely to be stimulated by the provision of cut and damaged areas of cellulose, and more easy access will be given to the susceptible S_2 layer. (With some fungi enzymes diffuse through the S_3 layer in order to attack the S_2 preferentially; in others S_3 has to be removed before attack on S_2 is possible). Latham *et al* (1978a, b) have shown that two types of rumen bacteria adhere preferentially to cut ends of cells in grass (and also to damaged surfaces): few organisms adhere to undamaged areas of cell wall. These organisms erode areas of substrate immediately surrounding them by cell-bound enzymes, and without adhesion of bacterium to substrate there would be little or no breakdown. In this case, cross-wise cutting of cells, although giving little percentage increase in total area, greatly increases the area of susceptible substrate available.

(iv) PARTICLE SIZE

Humphrey *et al,* (1977) grew a *Thermoactinomyces sp.* on three different particle sizes of Avicel; they found slight increases in the rate of growth and cellulose utilisation, with decrease in particle size, but more significant differences in total growth and cellulose utilisation. For the three size fractions tested, after 24 hours' growth the percentage cellulose digestion was 72, 80 and 89%. It is probable that since Avicel, a regenerated cellulose, is a porous relatively homogeneous substrate, there may not have been a very great deal of difference in surface area between the three particle size fractions. Overall rates would then be influenced mostly by mass transfer within the particles. The greatest change was observed when inoculum size was increased. Humphrey *et al,* (1977) used a shrinking sphere model with some success to represent the fermentation of Avicel particles with decreasing rate during the latter stages of fermentation when the growth was surface area limited.

Mandels *et al* (1971) have shown that for different particle sizes of cellulose, V_{max} for cellulolysis remains fairly constant while K_m

decreases with decreasing particle size. This is a similar effect to that noted by Moo-Young *et al* (1971) for growth of yeast on hydrocarbons; for increased rate of agitation in the reactor, which led to decreased particle (drop) size, μ_{max} remained constant while K_m decreased. This is typical of an apparent K_m which is controlled by diffusional transport where $K_m \alpha \sqrt{d_p}$ for spherical particles.

Clearly care must be taken in interpreting results concerning the influence of surface area on hydrolysis/growth rates.

(v) GRINDING AND BALL MILLING

Various methods of pretreating cellulosic substrates by grinding to reduce particle size have been examined and generally have been shown to increase the rate and extent of substrate hydrolysis (Millet *et al,* 1975; Dunlap *et al,* 1976; Mandels *et al,* 1971, 1974; Nesse *et al,* 1977; Polcin and Bezuck, 1977), microbial growth (Han and Callihan, 1974) and cellulase production (Greaves, 1971). Ball milling seems to be the most effective treatment of this type; it is held by some to reduce substrate crystallinity as well as particle size (Mandels *et al,* 1974).

4.4.2c Chemical factors

Cellulose is in nature usually associated and possibly complexed with other polymers, such as hemicelluloses and lignins. The presence of lignin decreases the susceptibility of cellulose to enzymic attack. Crawford (1974) has shown that in three wood pulps of varying lignin content the total weight loss of ligno-cellulose was inversely related to the initial lignin content. Similar results are given by other workers (Millet *et al,* 1975). It is interesting that a greater degree of delignification is required to improve digestibility of soft wood compared to hard wood, although the initial lignin content is similar in amount (Millet *et al,* 1975). Thus it is probably accounted for by qualitative differences in the lignins. Many methods of pretreatment have been devised to remove or alter lignin prior to cellulolysis and to swell the pore structure, although some have other effects like removing hemicelluloses (Millet *et al,* 1975; Dunlap *et al,* 1976; Han and Callihan, 1974; Mandels *et al,* 1974; Nesse *et al,* 1977; Rogers *et al,* 1972; Moo-Young *et al,* 1978; Peitersen, 1975a, b).

The methods of treatment include the following: sodium hydroxide, ammonia, concentrated and dilute acids, kraft process, acid bisulphite pulping, sulphur dioxide, peracetic acid, dry heat, steaming (under pressure), and irradiation. They have been used singly or in combination with each other or with milling. The most commonly used method is alkali treatment. This causes swelling of cellulose, decrease in crystallinity and disruption of the lignin structure. Alkali treatment greatly increases the value of agricultural and forestry wastes as a feed for ruminants and has been practised for nearly 60 years.

4.4.2d Other solid substrates

Less information is available on the pretreatment of other solid substrates prior to microbial growth or enzymic hydrolysis. Martin and So (1969) found that none of the many isolates of *Myxobacteria* that they tested would grow on native keratin as found in wool and feathers; however, they would grow on, and their enzymes would hydrolyse, keratin that had been modified by autoclaving. Noval and Nickerson (1959) showed that a strain of *Streptomyces fradiae* was able to decompose native keratin. Keratin is an insoluble, fibrous, structural protein with a highly ordered structure and is much less degradable in the native form than after it has been modified in ways that disrupt this structure. Heating generally has this effect; other treatments involving the reduction of the disulphide bridges (which hold the separate peptide chains together) are also highly effective.

Similarly collagen, another insoluble fibrous structural protein, is highly resistant to proteolysis. This resistance can be overcome by boiling which disrupts the collagen structure and converts it to soluble gelatin, which is easily degraded (Seifter and Harper, 1970).

Granular starch needs some pretreatment before hydrolysis with amylases (Dunlap *et al*, 1976) and it seems likely that most other insoluble substrates might benefit from some pretreatment - even if only milling to increase surface area.

4.4.3 Organism influence

Introduction

Several groups of microbes have the ability to degrade polymers: in this section of the review we are particularly interested in the relationships between the organism, its enzymes, and the substrate, and also in control of hydrolysis. It is convenient to split the problem into several sections, which are somewhat arbitrary but do show some basic trends. Accordingly information on the degradation of all polymers by *Myxobacteria* and *Cytophagas* has been put together. The next two sections deal with bacterial growth on cellulosic materials, the first considering aerobic bacteria and the second anaerobic organisms from the rumen. A short general section on fungal growth on cellulose is followed by more specific information on types of wood rot. The short sections on lignin and degradation of other polymers reflect the relative lack of information available.

4.4.3a Myxobacteria and Cytophaga

It is well known that *Myxobacteria* in general (Dworkin, 1966) and members of the *Cytophaga* group in particular are associated in nature with solid substrates (Stanier, 1942; Christensen, 1977). Stanier has shown that intimate contact is required for *Cytophaga* to degrade cellulose, and this may also be the case with other solid substrates Christensen 1977). The local erosion of cellulose fibres in the region of attached cells (which were regularly oriented in relation to fibre structure) was noted by Stanier (1942) who also observed that certain strains of *Cytophaga* produced copious slime during growth on cellulose. These findings were confirmed for *Sporocytophaga myxococcoides* by the electron micrographs of Berg *et al* (1972). This organism grew both on and inside the fibres. Several workers have found CMC degrading enzymes in culture filtrates of cytophaga (Berg *et al,* 1972; Marshall, 1973; Osmundsvag and Goksoyr, 1975) but none could detect Avicel or fibre degrading activity. Similarly, Chang and Thayer (1977) found that with their *Cytophaga sp.* cellulase was entirely cell-bound. It seems that cellulose fibre degrading enzymes are cell-bound, and for this reason *Cytophaga* need very close contact in order to degrade the substrate.

Chitin

Chitin, a highly insoluble β–(1–4) linked polymer of N-acetyl glucosamine is a common material in nature, being found in many fungal cell walls and in exoskeletons of many invertebrate animals. Sundarraj and Bhat (1972) studying chitin degradation found that there were two different types of strain of *C. johnsonii.* Strain C35 produced no clearing zones on chitin agar and chitinase activity was not found in liquid culture filtrates, although some activity could be found in sonicated cell free extracts. This strain grew only in static and not in shaken cultures, and the organism was found to be adhering to chitin particles in suspension. Growth in static culture was slow, probably because of oxygen limitation, but at about the same speed as Strain C31 in these conditions. However, C31 grew much faster (4X) in shaken than static conditions and some (although feeble) chitinolytic activity could be found in its culture filtrates; also it produced clearing zones on chitin agar. With both strains monomers of chitin were found to accumulate in the medium to a considerable extent during growth.

When a *Cytophaga sp.* was grown on algal galactans, Turvey and Christison (1967) found that galactan sulphate degrading activity was largely (c. 75%) cell-bound. In a later study it was found (Duckworth and Turvey, 1969) that agarase activity in this organism was associated with the polysaccharide slime layer; purification of the enzyme was difficult as it was found to be stabilised by the slime.

Kurowski and Dunleary (1976) examined pectinase production by *C. johnsonii;* this inducible enzyme could be cell-free or cell-bound (superficially located). In batch culture the proportion of cell-free enzyme started at less than 10% and increased so that at the end of active growth it was 40%; it continued to rise but only reached 100% long after the cessation of growth. At low dilution rate in continuous culture 90% of pectinase activity is cell-free while at high dilution rate only 35 to 45% is.

Christison and Martin (1971) found that a *Cytophaga sp.* grew in close contact with autoclaved feathers (largely keratin). Removal of the bacteria from the culture reduced the total protease activity by 50%, and the culture filtrate was found to have very poor feather solubilising activity compared to entire bacterial cultures. The protease which was basic was strongly associated with the acidic slime layer, which stabilised with it. The slime layer also functions to keep the bacteria and the enzymes in close contact with their substrate.

Some microbes can utilise the live or dead bodies of other microbes for their nutrient; this topic is reviewed by Stolp and Starr (1965). To do this they must first bring about lysis of the cell which can be achieved by means of certain antibiotics, other lytic agents or by cell wall destroying enzymes. Prey microbes can be viewed as mixed polymeric solid substrates, and the enzymes causing microbiolysis can in some cases be found extra-cellularly (Dworkin, 1966; Stolp and Starr, 1965). However, Shilo (1970) showed that for a *Myxobacter sp.* to lyse blue green algae close contact was essential, and agitation of the culture inhibited growth, probably because of interference with the host-parasite attachment. No diffusable lytic factor could be demonstrated so presumably the requisite enzymes are cell surface-bound. Similarly physical attachment of the parasitic bacterium *Bdellovibrio* to its host bacterium is essential (Stolp and Starr, 1965; Starr and Seider, 1971), although in this case the parasite may break through its host cell wall by physical rather than chemical (enzymic) means.

4.4.3b Growth of other aerobic bacteria on cellulose

While growing on cellulose *Cellvibrio fulvus* does not seem to attach to the outside of fibres but is found in large numbers in their lumens, thus the fibre is degraded from within. This difference from the *Cytophagas* may be due to the organism's flagellation and its inability to produce an adhesive slime. There are conflicting reports as to the location of cellulases during growth of *C. fulvus,* and it probably depends on the exact growth substrate (Berg *et al,* 1972 a, b; Berg, 1975), 'CMCase' seems to be predominantly cell-bound during growth on glucose and cellobiose, and possibly cell-free during growth on cellulose. Glucose represses cellulase synthesis, and CMCase is only produced (at low activity) when nearly all glucose in the medium has been used. 'Avicellase' activity seems to be half cell-free and half cell-bound. There is a definite tendency towards the release of enzymes at the end of active growth. Reducing sugars cannot be found in the medium during growth of *C. fulvus* on cellulose, suggesting that the substrate is not hydrolysed any faster than the organisms' ability to take up the degradation products (Berg *et al,* 1972 a, b; Berg, 1975). *C. gilvus* only produces cellulase during growth on cellulose and CMC, and not on glucose or cellobiose; also it grew better on the former than on the latter substrates (Breuil and Kushner, 1976). Physical contact between the bacterium and the cellulose was shown to be necessary for growth and induction of cellulase.

Yamane *et al* (1970) have shown that *Pseudomonas fluorescens var. cellulosa* produces three separate cellulases of differing substrate specificity. One of these enzymes is cell-bounded (wall associated) while the others are extracellular (Yamane *et al,* 1971). The proportion cell-free to cell-bound enzyme depends on the growth substrate, Suzuki (1975) considers this cellulase to be constitutive, its formation being controlled by catabolite repression.

Beguin *et al* (1977) examined cellulose-bound and free (but not cell-bound) cellulase activity in cultures of *Cellulomonas flavigena.* During active growth the former predominated, but as growth slowed and ceased it decreased and the latter increased and became predominant. This was probably caused by the utilisation of all the susceptible cellulose, which left the enzyme with no substrate to bind to; this seems to be a common occurrence with exo-enzymes. In *C. flavigena* cellulase is inducible (cellulose being the best inducer) and is also subject to catabolite repression by rapidly metabolised substrates (Beguin *et al,* 1977, Stewart and Leatherwood, 1976). A stable mutant in which cellulase was not catabolite repressible was isolated by Stewart and Leatherwood (1976). In this mutant cellulase could be induced by cellulose, cellobiose or sophorose; in the presence of an inducer glucose increased rather than repressed cellulase production, presumably owing to stimulation of growth. Knapp (unpublished information) has observed that colonies of *C. flavigena* growing on Walseth cellulose agar produced no clearing zones. Mutants could be isolated that grew equally well but produced clearing zones. Neither wild type nor mutants produced detectible cellulase (filter paper activity) in liquid culture on cellulose. This is probably the result of the enzyme being bound to cell or substrate. It is possible that in the mutants the affinity of the cellulase for the bacterium was reduced.

Two bacteria were isolated by Thayer (1978) from termite gut, and they grew well on mesquite wood. Their CMCase activity was assayed in culture filtrates, whole and broken cells. In a strain of *Bacillus cereus* most activity was cell-free although a substantial part was cell-associated, while in a strain of *Serratia marcescens* all activity was associated with broken cells.

Skinner (1972) found that mixed bacterial cultures from activated sludge grew in close contact with cellulose fibres ensheathing them and causing erosion.

4.4.3c Rumen bacteria

It has long been known that many bacteria in the rumen are attached to solid particles, which they cause to be considerably etched. Some of these attached bacteria are cellulolytic (others may degrade other polymers) (Hungate, 1966). It seems likely that those organisms which are attached to their substrate may employ cell-bound enzymes. King (1959) showed that a considerable amount of the cellulase activity found in rumen contents was associated with cell surfaces and other particles. Other workers (Krishnamurti and Kitts, 1969) found that no cellulase activity could be detected in cell-free samples of rumen fluid, and only a little in sonic extracts of microbes suspended in the fluid. Sonic extracts of organisms separated from 'rumen solid fraction' contained more activity, though not a great deal. No attempt was made to measure the cellulolytic activity of whole cells or cell envelopes. Electron micrographs (Akin, 1976) have shown that cocci and bacilli are attached very closely to the surface of leaves undergoing degradation in the rumen. Erosion of the solid substratum is much more pronounced in the close vicinity of the bacteria. He found that many cocci appeared to have capsules which anchored them to surfaces. This was not true of the bacilli, which were, however, still in very close contact with the solid surface. These bacilli often had irregular surfaces which mirrored the shape of the surface they were attached to. This may be an artefact of preparation; Akin thinks not. If he is correct, this organism shows an interesting adaptation to growth on solid substrates, resembling that of *Cytophaga* species, which also lack a rigid cell wall.

Ruminococcus albus adheres to cellulose fibres when these are added to a culture medium, Patterson *et al* (1975) consider that the adhesion is mediated by the cell's polysaccharide coat material. It has been reported, however (Smith *et al,* 1973), that most of this organism's cellulolytic activity is extracellular, being mainly substrate bound. The benefit resulting from the attachment of this organism to its substrate is probably the reduction of the transport barrier for products of enzyme degradation. Cellulase synthesis by *R. albus* is subject to catabolite repression by cellobiose (Fusee and Leatherwood, 1972) but the enzyme does not suffer end-product inhibition by this metabolite. Similarly, it has been shown by Latham *et al* (1978a) that *R. flavefaciens* adheres to the cell walls of grass leaves by means of its glycoprotein coat. The bacteria are particularly associated with cut ends of cells and damaged areas of cell surface, and obvious degradation occurs only immediately around attached bacteria. In this organism most cellulases and hemicellulases are associated with the cell surfaces of intact bacteria. Another rumen microbe, *Bacteriodes succinogenes,* grown on the same substrate, also adhered to solid material and similarly showed preference for cut edges of cells and damaged surfaces. (Latham *et al,* 1978b). Both these organisms showed preference for certain types of grass cells. In mixed cultures *R. flavefaciens* was predominantly associated with some types of cells and *B. succinogenes* with others. Zones of digestion around both types of organisms were observed, and adhesion was clearly essential for cellulolysis. It has been reported (Hungate, 1947) that *B. succinogenes* has a firmly cell-bound extracellular cellulase. It is clear that available surface area is very important to these organisms so that they can attach to and degrade cellulose. However, it must be noted that there is a preferential adhesion on cut and damaged areas. Therefore the total surface area will not be important compared with the area of especially accessible regions, such as cut and damaged surfaces. Similarly *Cellvibrio fulvus* cannot grow on natural length cotton fibres but will grow on the same fibres if they are chopped up, allowing the bacteria access to the fibre lumen (Berg *et al,* 1972). It seems that the internal surface of the fibre is more susceptible than the external surface to this organism. The preference of these two rumen organisms for certain types of cells and surfaces is very interesting. It may be that organisms can attach to any surfaces, but only grow well on the preferred ones. Alternatively the preference may be due to selective adhesion to the preferred surfaces. The selective adhesion could be due to differences in the chemical nature of the different grass and bacterial cell surfaces overall or possibly to the specific binding of surface-bound bacterial enzymes to their substrates.

Using a mixed culture from a similar system to the rumen, viz. anaerobic sewage digesters, Khan (1977) examined cellulose conversion to methane. Cellulose degradation rate was decreased by a high rate of stirring in the vessel, which it was thought probably interfered with attachment of bacteria to the cellulose fibres.

4.4.3d Fungal growth on cellulose

For general information on the growth of fungi on cellulose and on cellulysis by isolate enzymes readers are referred to several reviews and symposia published over the past ten years (Hajny & Reese, 1969 Wilke, 1975 Gaden *et al,* 1976; Bailey, Enari and Linko, 1975; Ghose, 1977).

Despite the enormous literature on cellulolysis relatively little information seems to be available concerning the physical aspects of growth and enzyme location in the better known and most studied fungi.

Trichoderma viride (IMI 92027) has been shown to grow in close contact with cellulose fibres, hyphae growing within the fibre lumen and inside fibre walls (Berg and Hofsten, 1976). Cellulase (CMCase and avicelase) is particle bound during growth of this strain, significant amounts of enzyme only appearing in the culture filtrate at stationary phase (Berg and Pettersson, 1977). Similar, though not so pronounced, results are often seen with *T. viride* mutants QM 9123 and 9414 in which most cellulase is not released until cessation of growth (Herrera-Zeppelin, person comm., Peitersen, 1975a & b; Moo-Young *et al,* 1977; Andreotti *et al,* 1977). These differences in degree probably reflect the difference between wild type and mutants. Berg (1978) also examined cellulase location in *Phialophora malorum* finding enzyme to be mostly cell-free; synthesis of cellulase was repressed by glucose and cellobiose.

It is well known that the highly cellulolytic fungus *Myrothecium verrucaria* and others, like *Chaetonium globosum* and *Pestalotiopsis westerdijkii,* grow well on cellulose but do not produce highly cellulolytic culture filtrates (Mandels, 1969, 1975; Reese, 1972). Probably either all or part of the cellulase complex is cell or substrate bound not being released at the cessation of growth. Which of these possibilities is correct is not known as it appears that enzyme location has not been studied in these systems.

4.4.4 Wood rotting

4.4.4a Substrate

Wood contains several different kinds of polymers, chief of which is cellulose. Wood rotting, being of economic importance, has been extensively studied and well reviewed (Liese, 1970; Norkrans, 1967; Jurasek *et al,* 1967; Nilsoon, 1973; Rossell *et al,* 1973). Bacteria are thought to play a relatively minor role in the disintegration of wood, compared to fungi. They are more prevalent than usual in very wet conditions (Rossell *et al,* 1973), fungi preferring a low moisture content.

To understand fungal decay of wood, it is necessary to appreciate the complex structure of the substrate. For a description of this readers are referred to more authoritative sources such as Jane (1970), Meylan and Butterfield (1972), Jurasek *et al* (1967) and Wilcox (1968). Wood cells differ in structure and composition and thus in susceptibility to microbial attack. In addition to this the separate layers within one type of wood cell differ from each other chemically. Fig. 4.1 represents in simplified form the structure of a woody cell. It must be pointed out that gross differences exist between hard and soft woods as well as between species of plant.

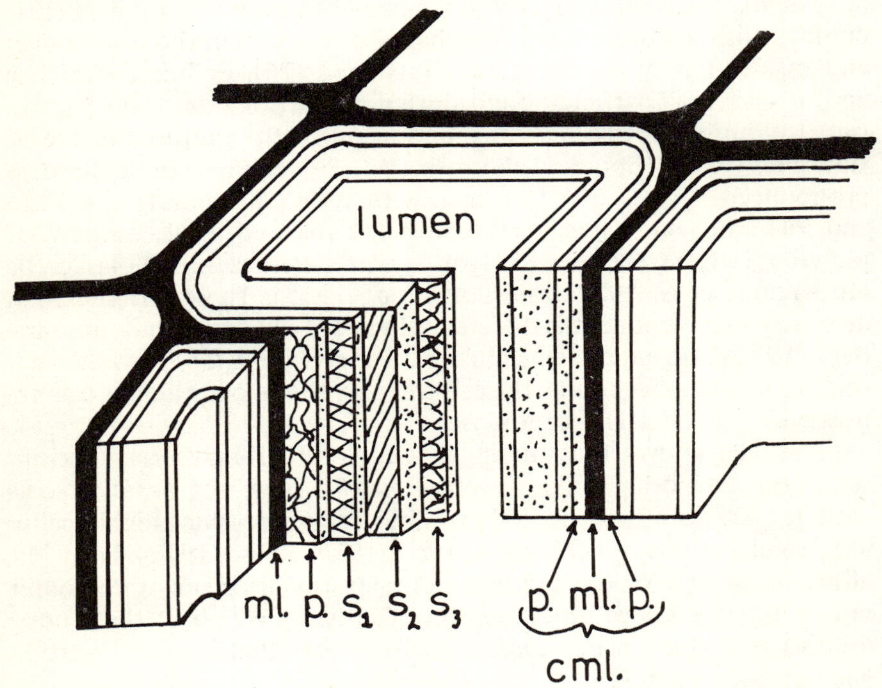

Fig. 4.1
Diagram of woody cell (in transverse and partial longitudinal section) to show the different layers of the cell wall. Note the differing orientation of cellulose microfibrils in the different layers of the wall. Middle lamella – ML., primary wall – p., compound middle lamella (i.e. true ml + two adjacent primary walls) – cml. The secondary wall – $s_{,1}$has three layers the outer – S_1, the middle – S_2 and the inner – S_3.

4.4.4b Organism

Fungi are the main decomposers of wood tissues; their growth on these substrates is well documented and has been classified. Wood rot has been split into five categories (Liese, 1970; Norkrans, 1967;

Jurasek *et al,* 1967), in addition to which are the litter decomposers. These classifications are based on chemical and physical effects rather than taxonomy of the causative organism, although certain taxa are associated with certain types of rot.

(i) *Blue-stain fungi* These are a group of Fungi Imperfecti and Ascomycetes which cause little decomposition and have little effect on the wood's physical properties. The blue staining is due to the colour of the hyphae showing through the wood cells.

(ii) *Soft rot* The causative organisms of soft rot are also a group of Fungi Imperfecti and Ascomycetes. This group, however, do cause considerable degradation of woody tissue. Nilsson (1973) and Corbett (1965) have examined soft rots as they affect both hard and soft woods, and distinguish two types of attack by these fungi. In type 1 attack, hyphae invade the tracheid lumens, via bore holes, align themselves with the long axis of the cell and produce side branches, which then grow parallel to the cell's long axis inside the S_2 layer, causing chains of cavities, of characteristic shape, to form within the tracheid wall, while the S_3 layer is not attacked (Nilsson, 1973). No cavities are formed in type 2 attack; hyphae after entering the cell lumen, lie against the S_3 layer of the wall and secrete enzymes which cause erosion of the cell wall in a 'valley' around the hyphae; the S_2, and sometimes the S_3 (especially in hard woods), layers are dissolved. Nilsson (1973) discusses the role of cellulases in relation to soft rot and considers that some fungi producing type 1 decay may have cell-bound enzymes, while type 2 rotters all produce cell-free diffusing cellulases. Some fungi can produce both type 1 and type 2 decay. Soft woods are less susceptible to soft rot than are hard woods, possibly owing to the greater degree of lignification. Cellulose and also hemicelluloses are degraded and there is some modification (probably demethoxylation) of lignin. (Nilsson, 1975; Norkrans, 1967). Within the cell the S_2 is always more susceptible to dissolution than the S_3 layer. Nilsson (1975) has observed that growth of some type 1 soft rotters on pure cellulose depends greatly on moisture level, fairly dry conditions being preferred.

(iii) *Brown rot* This is a group of Basidiomycetes which preferentially attack soft woods. Brown rotters decompose only the cell wall polysaccharides, lignin remaining virtually undegraded though somewhat modified by demethoxylation (Kirk, 1971; Liese, 1970; Norkrans, 1967; Jurasek *et al,* 1967). Hyphae, growing in the cell lumen secrete polysaccharases which diffuse through the S_3 layer and proceed to decompose the S_2 and S_1 layers before S_3 is degraded (although sometimes S_2 and S_3 are attacked at the same time); the highly lignified compound middle lamella is not attacked (Wilcox, 1968). Hyphae are seldom found inside the cell wall although they will grow into cracks and fissures in the wall if they develop.

(iv) *White rot* White rot fungi are a group of Basidiomycetes, which preferrentially degrade lignins and hemicelluloses, only degrading cellulose at a later stage. Hyphae grow inside the cell lumen after having penetrated via bore holes and enlarged pit canals. The layers of the cell wall are removed in order, S_3 first followed by S_2, S_1 and finally the compound middle lamella. However, cell wall thickness remains fairly constant at the onset of attack. Unlike brown rot cellulases the enzymes produced by white rotters cannot degrade lignified cellulose until lignin has been removed and also cannot diffuse through lignified areas of the cell wall to degrade less lignified areas. Thus cellulose is not degraded until the lignins and hemicelluloses have been removed (Wilcox, 1968). It is probable that lignin degradation yields little or no energy and may consume it (Kirk *et al,* 1976). White rot fungi show a preference for hard over soft woods.

(v) *Simultaneous rot* This group of fungi, preferentially attacking hard woods, are characterised by the fact that all the components of lignified cell walls are decomposed at the same rate (Liese, 1970; Jurasek *et al,* 1967). Hyphae enter cell lumens via bore holes, the layers of the cell wall are degraded successively, and there is gradual thinning of the wall during decay. Hyphae ramify across the cell surface, causing pronounced local lysis, such that they appear to lie in valleys. Cavities may appear inside the cell wall probably owing to enzyme diffusion.

4.4.4c Comparisons of different wood rot

In a beautifully illustrated paper Wilcox (1968) demonstrates the successive changes both physical and chemical brought about in wood structure by representative white and brown rot fungi on typical hard and soft woods. Nilsson (1974) found considerable differences in the ability in the two types of fungi to produce cellulases on test media. All the white rotters tested (31) produced clearing zones, while only seven out of 31 brown rotters did so. He thought that brown rot fungi might only produce C_1-cellulase while growing on lignified cellulose, which might be a preferred substrate to pure cellulose. Highley (1973, 1975) has shown that certain white rot fungi can degrade pure cellulose in the absence of wood, while certain brown rotters cannot: for the latter to degrade pure cellulose the presence of other substrates like wood, hemicelluloses, or even glucose is essential. Cellulase (C_1 and C_x) in the white rot fungi tested were shown to be inducible and their synthesis was subject to repression by simple sugars. However, the brown rotters examined produced only C_x cellulase which was constitutive; cellulase synthesis, far from being repressed, was increased by the presence of simple sugars and non-cellulosic polymers (Highley, 1973).

In the same paper Highley looked at the ability of hard and soft woods to induce cellulases in brown and white rot fungi, but found no significant differences which would account for host preference. In a later study (Highley, 1976) he compared hemicellulases of a brown and a white rot fungus, grown on hard and soft woods, finding very small differences which did coincide with the host preference and known hemicellulase content of the woods. Keilich *et al* (1970) examined the cellulases and hemicellulases of brown and soft rot fungi but found no differences to account for the different patterns of rot caused by the two types of fungi.

4.4.5 Other polymeric substrates

4.4.5a Enzyme location

The location of agar hydrolysing enzymes of *Pseudomonas atlantica* has been intesively studied by Yaphe and co-workers (Day *et al,* 1975. Day and Yaphe, 1975; Groleau and Yaphe, 1977) who have shown that sevaral enzymes are cell-bound or associated. These enzymes are not cytoplasmic, all being located outside the cytoplasmic membrane, probably bound to the cell wall or outer membrane. In a Gram negative bacterium isolated by Hofsten and Malmquist (1974) agarase activity was mainly cell-bound during growth, being released at stationary phase. Similarly, Reese *et al* (1973), screening many fungi for β–D–Xylopyranosidase activity, found that during early growth enzyme activity was associated with the mycelium, but later on most activity was found free in the medium. Pullulanase (α–1,6 glucanase) in a *Klebsiella* has been shown to be cell-bound and is associated with the outer membrane (Wöhner and Wöber, 1978). Lilley and Bull (1974) have shown that the β–1,3 glucanase of a thermophillic streptomycete is truly extracellular, as is the pectate lyase of *Hypomyces solani fsp. curcubitae* (Hancock *et al* 1970). Similarly, when *Rhizoctonia solani* is grown on sodium polypectate and CMC the appropriate hydrolases are mainly extracellular (Lisker *et al,* 1975).

Few authors give details of the location of microbial proteases. A *Halobacterium sp.* isolated by Norberg and Hofsten (1969) produces a cell-free extracellular protease. However, the enzyme does not appear in the medium till fairly late in growth. Corpe and Winters (1972) examined several hydrolytic enzymes found in a few marine strains of *Pseudomonas.* No clear pattern was established for other enzymes, but protease activity was mainly cell-bound, largely being associated with the cell envelope in five of six organisms tested. About 50% of the specific protease activity of the cell envelope could be removed by three washes with sea water, indicating a fairly weak binding. The organisms were Gram negative and this finding with those of Yaphe and co-workers agrees with the theory of Costerton *et al* (1974) on the cell-bound nature of hydrolysis enzymes in water borne microbes.

In *Lactobacillus bulgaricus* (one of the yoghurt fermentation organisms) Argyle *et al* (1976) have shown that all the proteinase is associated with the cell surface. In another food microorganism, *Mucor hiemalis* (used in the manufacture of Chinese cheese sufu), Wang (1967) has found very low yield of proteolytic enzymes in the medium, most activity being cell surface bound. The enzyme could be released by elution with sodium chloride, or by growing the organism at high salts concentration, suggesting that the enzyme was probably bound to the cell by weak ionic bonds.

4.4.5b Lignin

Reviews of the microbial degradation and transformation of lignin (Kirk, 1971, 1975; Ogelsby *et al,* 1967) show that while many organisms in several groups can cause minor alteration to lignin molecules or grow on the degradation products of lignin, only one group of organisms appears to be able to decompose and assimilate lignin completely. The group is the white rotting and litter-decomposing fungi in the Basidiomycetes (Kirk, 1971). Extracellular phenol oxidising enzymes can be found during growth of white rotters on lignin although it is possible that some lignin depolymerising enzymes are hyphae bound (Kirk, 1975). It is of interest to note that most of the bonds found in lignin are not amenable to hydrolysis; lignin therefore differs from most biological polymers in that depolymerisation is not hydrolytic and may require co-enzymes. Lack of suitable methods for the preparation or assay of lignin have somewhat hampered research on its degradation, but new methods of ^{14}C labelling of lignin have allowed recent studies of turnover of lignin in the natural environment (Hackett *et al,* 1977, Crawford *et al,* 1977 a,b). Comparison of the turnover rates of lignin and cellulose in specific woods by soil microbes shows that the latter is degraded 4 to 10 times more rapidly. Even with white rotters lignin degradation can be somewhat difficult. Kirk *et al* (1976) have shown that with two such fungi lignin cannot be degraded or serve as a growth substrate in the absence of other substrates such as cellulose and glucose. Kirk *et al* (1978) have also examined the influence of culture parameters on lignin metabolism by *Phanaerochete chryosporium.* The degradation of 5 mg of lignin by this fungus required 100 mg of glucose. Agitation of the culture was found to suppress lignin metabolism, so the fungus was grown in static culture. Under these conditions the oxygen concentration in the atmosphere above the culture was very important. At 5% O_2 there was no degradation, while at 100% O_2 degradation was 2 to 3 times as fast as in air.

4.5 REGULATIONS OF EXTRACELLULAR METABOLISM

4.5.1 Enzyme synthesis

If it is assumed that the production of extracellular enzymes required to degrade polymers outside the cytoplasmic membrane, is controlled in the same way as the production of intracellular enzymes (Watson, 1977) then it is reasonable to assume that the polymers themselves do not serve directly as effectors of enzyme synthesis - as they cannot enter the cell.

(i) *Induction* The cellulase complex *T. reesei* (formerly known as *T. viride*) is known to be inducible (Mandels and Weber, 1969) and is induced in the presence of cellulose (which cannot be the intracellular inducer) and also cellobiose and several simple saccharides. Cellobiose is the major product of cellulase and it is likely that it (possibly together with small cellodextrins) is the true inducer. Other polysaccharases are reduced by related oligomers (Bull, 1972; Reese 1972). The chitinase of *Serratia marcescens* is best induced by chitin, although it can be fairly well induced by chitobiose but not n-acetyl glucosamine (the monomer). The most likely intracellular inducers are small soluble chitodextrins (Monreal & Reese, 1969). Chitinase is repressed by small rapidly metabolised compounds like glucose. Groleau and Yaphe (1977) showed that several enzymes involved in agar degradation by *Pseudomonas atlantica* were induced by agar, neoagarobiose and neoagarotetraose. Similar examples are given by Pollock (1962). Even for soluble substrates, such as amino acids, degradative enzymes are often induced by degradation products (Ornston, 1971) thus here, and presumably in the case of polysaccharases, induction requires the presence of a certain basal amount of enzymes to convert substrate to product-inducer. If an enzyme is a *catalyst* the reaction will *occur* however slowly in the absence of enzyme. Products thus formed (in very small concentrations) may be sufficient to induce the enzyme.

(II) *Repression T. reesei* cellulase can be produced during growth on commercial glucose as this contains low concentrations of sophorose, a very powerful inducer of cellulase. Even so, cellulase is not produced unless the growth rate is very low (in continuous culture) or until the glucose has been used by the organism (Brown *et al,* 1975), as glucose at high concentrations will decrease cellulase production in a fermentation because of catabolite repression. Cellobiose can give similar results, yet at low concentrations both these compounds can improve cellulase yields by stimulation of growth (Mandels and Weber, 1969). It seems that the concentration of catabolite repressors is not as important as the rate at which they are metabolised by the organism (Reese, 1972; Demain, 1972). Mutants of *T. reesei* which are hyper-producers of cellulase have been isolated (Mandels *et al,* 1971), and recently, with the aid of a new plate assay technique, mutants in which cellulase is not catabolite repressible have been isolated (Montenecourt and Eveleigh, 1977a, b).

Most catabolic extracellular enzymes seem to be subject to catabolite or end product feedback repression (Bull, 1972; Demain, 1972). Some such enzymes seem to be constitutive and only under repressive control, as are the cellulases from *Myrothecium verrucaria* (Hulme and Stranks, 1971) and *Pseudomonas fluorescens var. cellulosa* (Suzuki, 1975) and some other β–glucanases (Santos *et al,* 1977). Many cellulases, however, appear to be under the dual control of induction and catabolite repression: this is true for *Verticillium albo-atrum* (Gupta and Heale, 1971), *Cellulomonas* (Stewart and Leatherwood, 1976, Beguin *et al,* 1977), *Cellvibrio gilvus* Breuil and Kushner, 1976), *Aspergillus awamori* (Enari *et al,* 1975), *Sporotrichum pulverulentum* (Eriksson, 1978) and some other white-rot fungi (Highley, 1973).

4.5.2 Co-metabolism

Some brown-rot fungi are a strange case, as it seems that cellulase synthesis was constitutive, and far from being repressed by them seemed to require the presence of more easily degraded carbon sources such as glucose or hemicellulose (Highley, 1973, 1977), cf. the degradation of lignin (Section 3.4b). It is possible that in these cases degradation of the polymer in question cannot produce enough energy to allow induction of the enzymes and growth of the organism. This phenomenon, known as co-metabolism, often occurs with pesticides and other recalcitrant pollutants, which will not support growth but can be degraded in the presence of an energy source (Horvath, 1972).

4.5.3 Enzyme release

It often happens that extracellular enzymes only appear in the medium late in the growth of a culture. There are several possible reasons for this and often all of them are not considered by investigators. During active growth cellulases (and other enzymes) are often cell- or substrate bound, only appearing in the medium as the growth rate slows down. This is probably associated with a decrease in the area of substrate available for adsorption of enzymes and/or cells. Thus at the end of growth enzymes are desorbed from residual substrate and released from the surface of non-growing cells. Also if catabolite repressors are present in the medium cellulases may not be produced until the repressors are mostly degraded and the growth rate has slowed down, i.e. late in the growth of the culture.

Ghose *et al* (1975) grew *T. viride* on a cellobiose medium; they claimed that *de novo* synthesis of cellulase only starts at or near the cessation of growth, and that most enzyme appears after this time. To explain this they suggest that cellulose m-RNA is extremely stable and continue to produce the enzyme for several days after the cessation of growth. An alternative explanation of their results is that cellulase is partially cell-bound and is slowly desorbed from the mycelium after the cessation of growth. However, both these postulates are hard to test. Other cases in which enzymes are released very late in growth are known, and this is especially true for proteases and some other enzymes released by *Baccilli* (Keay *et al,* 1972). Release of these enzymes may or may not be associated with sporulation (Levisohn and Aronson, 1967). Coleman (1967) has studied regulation of exo-enzyme formation in *B. subtilis* and proposes that control is by the availability of nucleotides to make specific m-RNA molecules. Terui has modelled exo-enzyme formation by some microbes and accounts for prolonged enzyme production after cessation of growth on the basis of a long-lived m-RNA (for which he has some experimental evidence). (Terui, 1972; 1972; Bull, 1972).

Usually catabolite repression occurs at the level of transcription (Watson, 1977; Demain, 1972), but some workers claim that for some exo-enzymes in certain microbes, it occurs at the translation level (Terui, 1972; Nisizawa *et al,* 1972).

4.5.4 Enzyme inhibition

Enzymes (especially anabolic ones) are often inhibited by their own products, which can prevent wasteful overproduction. Although the exo-enzymes that we are dealing with are catabolic the same rule applies. If hydrolysis products build up outside the cell faster than the organism can use them they may be lost by dilution or encourage the growth of other microbes at the expense of the depolymerising organism. End product inhibition of preformed extracellular depolymerases (as well as catabolite repression of their synthesis) will ensure that the concentration of small molecules will not increase to too high a level, thus preventing wastage.

The cellulase complex of *T. reesei* is well known to be inhibited by cellobiose –its major end product. (Norkrans, 1967; Mandels and Weber, 1969; Mangat and Howell, 1978).

Halliwell and Griffin (1973) have shown that cellobiose is a competitive inhibitor of C_1-cellulase, while glucose is not. Product inhibition of cellulase has been modelled by Stuck and Howell (1975) and Mangat and Howell (1978). It has been shown that β-glucosidase. although it has no cellulolytic activity, causes greatly increased activity of purified C_1 and combined C_1 and C_x fractions of *Trichoderma* cellulase (Halliwell and Griffin, 1973; Wood, 1975). It is interesting that β-glucosidases from other organisms will increase total cellulolysis by the cellulase complex from organisms deficient in β-glucosidases. This synergism is due to the removal of product inhibition by cellobiose (Wood, 1975; Sternberg *et al,* 1977).

Synergism has also been noted in the growth of microbes. Often defined mixed cultures of cellulolytic and non-cellulolytic organisms give more or faster growth than the cellulolytic microbe in pure culture. This effect has been attributed to the removal of inhibitory intermediates of cellulolysis, like cellobiose (although in some cases other factors, such as cross-feeding, may be involved). Several workers investigating SCP production have found dual culture to be advantageous (Ek and Eriksson, 1975; Srinivasan, 1975; Dunlap, 1975; Smith and Neudoerffer, 1971, Edwards and Smith, 1975; Smith *et al,* 1975, Peitersen, 1975b). Similar observations have been made by other authors involving *Sporocytophaga* (Hofsten *et al,* 1971), *Clostridium thermocellum* with *Methanobacteria thermoautotrophicum* (Weimer and Zeikus, 1977), and a culture of three anaerobic bacteria - one of which was cellulolytic (Enebo, 1949). In the anaerobic cultures it is interesting that the fermentation pattern is completely changed in the mixed culture.

It is likely that many other exo-enzymes are subject to similar end product inhibition.

4.6 SINGLE-CELL PROTEIN FROM CELLULOSE

At the present time there are a number of operating plants in the world producing microbial protein from liquid substrates such as petroleum hydrocarbons, molasses, sulphite pulping liquor and starch wastes (Moo-Young, 1976). Detailed information on the processes and amounts of product is very sparse, and figures for fermenter productivities, which are generally commercial secrets, are not widely published. Some productivity values are quoted by Moo-Young (1977) for specific substrates. These are (in kg cells/m^3/hr = g/l/hr): Paraffin–3, Methanol–2, Ethanol–4.5, Molasses–5.2, Acetate–4.5.

Other figures appear elsewhere, namely: Methane–2.39 (Sheehan and Johnson, 1971), sulphite liquor–3.4. (Pekilo process, Romantschuk, 1975), Methanol–5.2 (Dostalek and Molin, 1975). Recently much interest has centred on production of SCP from renewable resources such as carbohydrate, especially in waste materials (Forage and Righelato, 1978) including cellulosic materials, which are potentially available in huge amounts as agricultural, forestry and domestic wastes (Bellamy, 1974; Humphrey, 1975).

Many cellulolytic micro-organisms are probably suitable for use as SCP - especially for animal feed, although few have been subjected to nutrition and toxicity tests as yet. Unfortunately, as has been mentioned previously most cellulose occurs in forms which are only slowly biodegraded. It is therefore likely that most of the cellulosic substrates available will need to be subjected to a pretreatment process to improve their degradability. These processes are expensive and will effectively increase the cost of the basic feedstock for SCP production. A second problem is that many cellulolytic organisms are intrinsically slow growing, and even when growing on pretreated cellulose, mean doubling times for cultures are relatively long, the shortest being of the order of 3 to 5 hrs for *Cellulomonas/Alcaligenes* growing on pretreated bagasse (Dunlap, 1975) or *ca* 4 hrs for *Thermoactinomyces sp.* (Nolan & Forro, 1976). Often doubling times are in the order of 20 hrs or longer (e.g. Peitersen, 1977; Ek and Eriksson, 1975). Growth curves are often linear rather than logarithmic, (e.g. Humphrey *et al,* 1977; Bellamy, 1974; Dangulis and Bone, 1977; Moo-Young *et al,* 1977; Peitersen, 1975a,b) suggesting that growth rates are limited by accessible surface area or by transport effects. This situation has certain parallels with hydrocarbon fermentations in which growth can be linear if there are mass transfer limitations. Growth rates can thus be increased by increasing the surface area accessible to organisms or enzymes.

Collected information on the growth of microbes on cellulosic substrates is given in Table 4.2. Of particular interest are the sections on productivity and yield factors - as these will greatly influence the economic viability of SCP production from cellulose. These figures should be contrasted with those given earlier for commercial processes. Readers will note that the earlier figures refer to biomass, and the figures for cellulose fermentations to protein production. However, even assuming only 25% protein (it may be 60% for bacteria) it can quickly be seen that the productivity of protein from cellulosic substrates is lamentably low. In almost all cases productivity from cellulose is at least one (and occasionally two) orders of magnitude below that of commercial processes using liquid (or gaseous) sub-

Table 4.2

Micro-organism	Substrate	Pretreatment	Conc. %w/v	Bio-mass g/l	Protein g/l	% protein in product	Culture time hrs	Productivity g protein/1/hr	Yield g protein /g substr.	Reference
T. viride (QM 9123)	Barley straw	5.7% NaOH and washed	1	–	1.84(N)	23.8	48	0.0383	0.184	*Peitersen (1975)
C. cellulolyticum	Wheat straw	Untreated	1	–	0.43 (N or Biuret)	4.6	48	0.0089	0.043	Chahal *et al* (1977)
C. cellulolyticum	Wheat straw	1% NaOH	1	–	2.51	40.6	36	0.0697	0.251	
	Wheat straw	1% NaOH/ peracetic acid	1	–	2.67	42.9	36	0.0741	0.267	
	Sigmacell	–	1	–	1.4	17.5	48	0.0291	0.14	
T. viride (QM 9414)	Sigmacell	–	1	–	0.8	13.5	48	0.0166	0.08	
Myrothecium verrucaria	Newspaper	Ball milled	4	–	3.3(N) 1.42(Biuret)	9.8	144	0.023 0.010	0.0825 0.0355	*Updegraff (1971)
M. verrucaria	Woodpulp	Ground to 60	1.5	–	1.49(N)	17.85	120	0.0124	0.1	*Chahal & Gray (1970)
Chaetomium globosum			1.5	–	1.2	10.3	120	0.01	0.08	
Rhizoctonia solani			1.5	–	1.86	20.3	120	0.0155	0.124	
Trichoderma sp.			1.5	–	1.49	23.8	120	0.0124	0.1	
Chaetomium cellulolyticum	Solkafloc Cellulose	–	1	–	2.7(Biuret)	40.3	36	0.075	0.27	Moo-Young *et al* (1977)
Trichoderma viride (QM 9414)	–	–			1.1	22.4	36	0.0306	0.11	
C. cellulolyticum	Saw dust	Ground to 40 mesh 1% NaOH			1.8	40	36	0.05	0.18	

T. viride (QM 9414)					1.3	25.4	36	0.036	0.13	
C. cellulolyticum	Hardwood sawdust	Hydrolysed wood: plus solubles	1							
		Mild acid	1		1.31	15	43	0.03	0.13	Moo-Young, Chahal & Vlatch (1978)
		Strong acid	1		1.38	21	132	0.043	0.138	
		Alkali	1		0.78	11	28	0.027	0.078	
T. viride	Newspaper		1		0.77(N)	11	150	0.005	0.077	Brown & Fitzpatrick (1976)
T. viride (QM 9123)	Barley straw	NaOH treated, then milled to 40 mesh	2	–	2.66(N)	c.21	144	0.0184	0.13	*Peitersen (1975)
T. viride and *C. utilis*			2	–	2.1	c.21	96	0.0218	0.105	
T. viride and *C. utilis*										
T. viride and *S. ceraviseae*			2		2.6	c.21	120	0.0217	0.13	
T. viride (QM 9414) (continuous)	Pure wood cellulose	Ball milled	1.1	4.0	2.4(Biuret)	32	D=0.038 RT= 26	0.091	0.22	*Peitersen (1977)
Cellalomonas sp. + *Alcaligenes sp.*	Rice straw (ground to 20 mesh	nil	1–5	–	0.6(Lowry)	–	120	0.005	–	*Hans & Callihan (1974)
		4% NaOH 100°Cg15 min	1–5	–	1.76	–	120	0.015		
	Bagasse (ground to 20 mesh)	nil	1–5	–	0.54	–	120	0.0045	–	
		1%NaOH Room Temperature	1–5	–	2.44	–	120	0.02		
		3%NaOH Room Temperature	1–5	–	4.64	–	120	0.039		
		3%Na, 260 psig 5 min.	1–5	–	7.2	–	120	0.06	–	
Cellulomonas sp. + *Alcaligenes* (continuous)	Bagasse	NaOH	?	10	c.5.0	50–55	RT = 10 D = 0.1	1(cells) 0.5(protein)	0.2 0.1	Dunlap (1975)

Table 4.2 *(c'td.)*

Micro-organism	Substrate	Pretreatment	Conc. %w/v	Bio-mass g/l	Protein g/l	% protein in product	Culture time hrs	Productivity g protein/1/hr	Yield g protein /g substr.	Reference
Cellulomonas sp.	Cellulose powder	NaOH	2	4.5	2.3	–	96	0.024	0.115	*Shrinivasan (1975)
Cellulomonas sp. + *C. guillermondii*			2	10.2	4.4	–	96	0.046	0.22	
Cellulomonas sp. + *T. cutaneum*			2	11.8	4.8	–	96	0.5	0.24	
Pseudomonas JM 127	Mesquite sawdust		7.5		6.3(Biuret)	11.1	72?	0.0875	0.084	*Thayer (1976)
Brevibacterium	Mesquite powder		1.		0.53(Biuret & Folin)	–	72	0.0074	0.053	*Fu & Thayer (1975)
Cytophaga johsonae	Mesquite powder		5		1.68(Folin) 2.43	8.91 9.83	24 72	0.07 0.034	0.0336 0.0486	*Chang & Thayer (1975)
Mixed bacteria	Newspaper	Fine ground	2	1.32	c.0.66	2.5(Biuret) 3.7(N)	96	c.0.007	c.0.033	*Paredes-Lopez & Gonzalez (1973)
Sporotrichum pulverulentum	Powdered cellulose	–	1	–	0.34(N)	4.3	168	0.002	0.034	*Eriksson & Larsson (1975)
		N source–NH_4^+	1	–	0.5	14	144	0.0035	0.05	
	Mechanical waste fibres	N source–urea	1	–	0.33	20	48	0.0069	0.033	

S. pulverulentum (semi-continuous)	Waste chemical fibres	–	2–3	7.10	1.9–2.7(N)	–	average R. T. 20 to 28 D = 0.05 to 0.035	0.095	c.0.09	*Ek & Eriksson (1975)
Phanerochaete chrysosporium	Maple bark	Ball milled NaOH extracted acid hydrolysed	4 2 1 0.5	– – –	2.3(Biuret) 1.9 1.7 1.4	– – – –	75 75 75 75	0.031 0.025 0.023 0.019	0.0575 0.095 0.17 0.28	*Daugulis & Bone (1977)
Thermactinomyces sp.	Micro-crystalline Cellulose Avicel	–	5 5	11 14	c.5.5 c.7.0	? ?	16 42	c.0.343 c.0.17	0.11 0.14	*Humphrey *et al* (1977)
Thermophilic actinomycete MJ	Avicel Swollen cellulose	– 85% H_3PO_4	0.5 0.5	– –	0.53(Lowry) 0.55	? ?	13 20	0.04 0.0275	0.106 0.11	*Su and Paulavicus (1976)
Thermactinomyces sp.	Cellulose		1		1.75	10.5	13	0.135	0.175	*Nolan & Forro(1976)
Thermononospora fuscu	Wood fibres (pulping fines)	–	0.5	–	0.4	30	96	0.004	0.08	*Crawford *et al* (1973)

Table 4.2 – Collected data on fermentations involving cellulosic substrates. Note that productivity and yield values refer to the amount of protein rather than total biomass; this is because it is often difficult to assess biomass when mixed with residual cellulose. Where available, information is given on the type of protein assay used, Biuret, Folin, Lowry (Folin method) or N (total Kjeldahl nitrogen X6. 25). References marked with an * are those in which the authors have not themselves worked out productivity, yield, etc. These have been evaluated by the reviewers from the information in the original article. The yield value refers to the initial weight of cellulosic substrate provided and not to the amount actually consumed – again this is for technical reasons.
RT = Retention time (hrs) D = Dilution rate (hrs^{-1})

strates. The only processes with relatively high productivity are those for growth of *Cellulomonas/Alcaligenes* on bagasse and *Thermoactinomyces* on cellulose.

The fact that these figures are very low does not necessarily mean that the processes would be uneconomical, but it does mean that the running costs for operating the fermenter (temperature control, stirring, aeration, maintenance of sterility, etc.) will be much higher for cellulosic fermentations than for many others.

Without detailed economic analysis it is only possible to say that it seems unlikely that any of these processes would be economic. There is a great deal of room for improvement in productivity, and it is to be regretted that many authors do not seem aware of the importance of yield and productivity and do not bother to evaluate them. Similarly protein content of the final product is often not determined even though it may be very low, e.g. 3 to 4%. Values for the yield of protein vary considerably, but are often around 0.1 to 0.2 (g protein/g substrate provided). This is quite reasonable, considering that yields from glucose would probably be around 0.2 to 0.25.

In several studies yield or productivity or both are improved by the use of defined mixed cultures of a cellulolytic and a non-cellulolytic microbe. This improvement may be due to the degradation of antagonistic chemicals or the rapid removal of degradation products of cellulose (cellobiose and glucose) which, if they accumulate in the culture, will repress the synthesis of more cellulase and inhibit the action of existing cellulase.

4.7 MATHEMATICAL MODELS

4.7.1 The substrate

(i) *Physical factors* The superficial substrate area accessible to the cell-free or cell-bound extracellular enzymes must be related to the degree of degradation. Stone *et al* (1969a) measured the accessible area after pretreatment and showed that initial degradation rates were proportional to this area. They did not model the change of accessible area with degree of degradation. Humphrey *et al* (1977) described a model which assumed the accessible area to be on the surface of spheres whose size and hence surface area shrank as fermentation proceeded. The overall model showed some similarity to the data obtained, but it was not determined how sensitive the results were to assumptions regarding the surface. Since initial mesh size of their Avicel particles did not relate to initial degradation rates, which were related rather to inoculum size, it is unlikely that the 'spheres' modelled could have been of a size commensurate with the overall particle dimension. Electron micrographs suggested a rough surface but not necessarily a micro-porous one. Other substrates such as wood have been shown to have an essentially porous structure. Changes in this porous structure during degradation were shown to occur by enlargement of pore size within the lumen, thus resulting in increased superficial area (Wilcox 1968).

Two alternative simple models are thus available for change of accessible surface with degradation.

a) Degradation of spheres from the outside
Substrate concentration C
Initial concentration C_0
Surface area S
Sphere radius R
Particle density ρ
Number of particles p.u.v. n

$$C = \frac{4}{3}\pi R^3 n\rho$$

$$S = 4\pi R^2 n = 4\pi n\left[\frac{3C}{4\pi n\rho}\right]^{2/3} .$$

If the reaction is first order in the exposed substrate concentration then

$$C = \left[C_0^{1/3} - \alpha t\right]^3 .$$

a) The second model considers the accessible surface to be the inside of cylindrical pores, and initially degradation can increase accessible area. For a pore of radius R length L if $R/L \ll 1$ and the reaction is first order in the substrate in the initial stages

$$C = C_0 - \kappa e^{kt} .$$

where k is a function of the local enzyme concentration and κ is a function of the porosity and pore size.

Particle size can be affected by ball milling and grinding, and the pore size can be altered by swelling or drying. Other pretreatment of cellulose, such as a caustic soda treatment, can also affect the chemical properties such as the degree of polymerisation. It is to be regretted that many studies on the effect of pretreatment have correlated pretreatment empirically with degradation rate. Unfortunately the parameters directly affecting the rate were not measured.

(ii) *Chemical factors* Early models of enzymatic cellulose degradation or saccharification (Van Dyke, 1972) regarded the division of cellulose into fractions of varying crystallinity as being crucial to the degradation pattern. It has been regarded as axiomatic that the decline in saccharification rates late in any given run on a crystalline material were due to the material becoming more crystalline and thus resistant to degradation. Direct observation (Humphrey *et al,* 1977) shows that using a *Thermoactiomyces sp.* on an Avicel substrate resulted in initial rapid hydrolysis but a later slowing down as hydrolysis only occurred from the edges and ends of the crystalline cellulose and not from smooth faces of the crystal. However, an elaborate model by Okazaki and Moo-Young (1979a) based on earlier models of Suga *et al* (1975) and Okazaki and Moo-Young (1978) has shown that initial low degradation rates of crystalline cellulose can be accounted for by their high degree of polymerization. Furthermore Suga *et al* (1975) show that as hydrolysis proceeds the distribution becomes weighted to lower and lower molecular weights rather than leaving substantial amounts of undegraded high DP cellulose behind. Humphrey *et al* (1977) suggest that although natural crystalline cellulose is degraded the crystals tend to lose imperfections during hydrolysis and become smoother surfaced, presumably limiting their capacity for adsorbing the hydrolysing enzyme and resisting further degradation.

Mangat and Howell (1978) and Howell and Stuck (1975) have shown that with a moderately crystalline cellulose, 'Solka floc', the initial decline in degradation rates can be modelled as a product inhibition effect, and later declines were shown by Howell and Mangat (1978) to be modelled by degradation of the enzyme already adsorbed to the substrate. Huang (1975), Lee *et al,* 1978 and Humphrey *et al,* (1977) among others have also found competitive product inhibition to be a dominant feature of the kinetics.

There are no quantitative models that have been found which describe the problem of transport of the polymer subunits from the solid to the cell surface and how this is affected by the adsorption of cells to the solid. The problem of defining the circumstances under which it is efficient for a cell to produce cell-free degradative enzymes is as yet unsolved and is not even clearly defined.

4.7.2 The kinetics of enzymic degradation of polymers

A polymer chain may be degraded by two distinct types of enzyme - exo-enzymes and endo-enzymes. The former cleave sub-units from the end of a chain (such as cellobiose for exo β 1,4 glucanase degrading cellulose). The latter cleave single bonds in the middle of a chain (such as the 1,4 linkage in cellulose by endo β 1,4 glucanase). The mode of action of endo- and exo-enzymes has been studied by Suga *et al,* (1975). Wheatley and Moo-Young (1977), Okazaki and Moo-Young (1978, 1979a,b), and Maxham and Maier (1978). In order that the essential features of their rather complex kinetic models can be clearer a simpler version is developed here.

(i) EXO-ENZYME ACTION

Assume that the enzyme acts by splitting monomer unit from the end of a chain and that the enzyme remains attached to that chain until it has been completely degraded. The reactions are then with a chain of length i etc.

$$E_1 + C_i \underset{K_{-1}}{\overset{K_{+1}}{\rightleftharpoons}} E_1 C_i \qquad \text{(C.1)}$$

$$\downarrow k_2$$

$$E_1 + C_{i-1} \underset{K_{-1}}{\overset{K_{+1}}{\rightleftharpoons}} E_1 C_{i-1} + C_1$$

$$\downarrow k_2$$

$$E_1 + C_1 \underset{K_{+1}}{\overset{K_{-1}}{\rightleftharpoons}} E_1 C_1 \quad . \qquad \text{(C.2)}$$

It is also assumed that the rate of degradation of the polymer chain is independent of length.

Normal enzyme kinetic assumptions are then used, namely that the enzyme-polymer complex concentrations $[E_1 C_1]$ are in a quasi steady state.

$$\mathrm{d}\frac{[E_1C_i]}{\mathrm{d}t} = 0 = k_2[E_1C_{i+1}] - k_2[E_1C_i] - k_{-1}[E_1C_i] \qquad \text{(C.3)}$$

Conservation of enzyme yields $E_{10} = E_1 + \sum_{i=1}^{N} E_1 C_i$

where the maximum chain length is N.

Equation (C.3) is summed over all chain lengths

$$k_{+1}[E_1] \sum_{i=1}^{N} C_i = k_{-1} \sum_{i=1}^{N} [E_1C_i] \quad .$$

If it is assumed that $[E_1\ C_1]$ is in equilibrium with $[E_1]$ and $[C_1]$

then $E_1 = \dfrac{E_{10}}{1 + /S_N/K_1}$ where $K_1 = \dfrac{K_{-1}}{K_{+1}}$

and $\dfrac{dC_1}{dt} = \dfrac{k_2\ E_{10}(S_N - C_i}{K_1 + S_N + C_1}$

where $S_N \equiv \sum_{i=1}^{N} C_i$ is the total number of polymer molecules present.

$$\frac{d\ S_N}{dt} = \frac{d\ C_1}{dt}$$

(ii) ENDO-ENZYME ACTION

The endo-enzyme is assumed to react with equal probability with any accessible bond. The rate of reaction is thus proportional to $(i-1)$ for a chain of length i.

The QSSA is also made in this case.

$$E_2 + C_i \underset{k'_{-1}}{\overset{k'_{+1}}{\rightleftharpoons}} E_2C_i \xrightarrow{k'_2} E_2 + C_{i-j} + C_j$$

for all i,j $j = 1, \ldots i-1, i = 2, \ldots N.$ for $i = 1$ only the first reaction occurs.

Now $E_{20} = E_2 + \sum_{i=1}^{N} E_2C_i$.

Thus $\dfrac{d}{dt}(E_2C_i) = 0 = k'_{+1}\ (E_{20} - \sum_{i=1}^{N} [E_2C_i])\ (i-1)\ [C_i]$

$$-(k'_{-1} + k_2')\ [E_2C_i] \quad .$$

Summing over all i

$$\sum_{i=1}^{N} E_2C_i = \frac{k'_{+1}\ E_{20}\ \Sigma(i-1)C_i + k'_2\ [E_2C_1]}{(k'_{-1} + k'_2) + \sum_{i=1}^{N} C_i\ (i-1)} \quad .$$

Now the total monomer molecular units present $S_w \equiv \Sigma_i C_i$

$$\therefore \Sigma(i-1)C_i = S_w - S_N \quad .$$

The rate of formation of monomer C_1 is

$$\frac{dC_1}{dt} = 2k_2' \sum_{i=2}^{N} \frac{E_2 C_i}{(i-1)} \quad .$$

The total number of molecules S_N is given by

$$\frac{dS_N}{dt} = k_2' \sum_{i=2}^{N} E_2 C_i \quad .$$

Noting that the summations in the last two equations exclude a term in E_2C_1 some manipulations is required to obtain finally

$$\frac{dC_1}{dt} = \frac{2k_2' E_{20}(S_N - C_1)}{K_S + (S_W - S_N) + KC_1} \qquad K_S = \frac{k_{+1}' + k_2'}{k_{-1}'} \qquad K = \frac{k_2'}{k_{-1}'}$$

$$\frac{d\,S_N}{dt} = \frac{k_2' E_{20}[(S_W - S_N) - C_1]}{K_S - (S_W - S_N) + KC_1} \quad ,$$

and the *combined* effect of both enzymes is then

$$\frac{d\,S_N}{dt} = \frac{k_2' E_{10}|(S_N - C_i)}{K_i + S_N} + \frac{k_2' E_{20}[S_W - S_N - C_1]}{K_S + (S_W - S_N) + K\,C_1}$$

$$\frac{d\,C_1}{dt} = \frac{2\,k_2' E_{20}\,(S_N - C_1)}{K_S + (S_W - S_N) + K\,C_1} + \frac{k_2' E_{10}\,(S_N - C_i)}{K_i + S_N} \quad .$$

These equations allow S_W to be a constant, assuming no fermentation is occurring. Fermentation would decrease both S_W and C_1 by the same rate term if it were assumed that the organism could only metabolise monomer.

It should be noted that in the development of the equation above product inhibition appears such as has been modelled by several workers including Howell and Stuck (1975), Howell and Mangat

(1978), Huang (1975) and Kim (1974). The effect above is competitive whereas Okazaki and Moo-Young (1978) assumed a non-competitive inhibition in developing their model of the effect of endo- and exo-enzymes. A synergistic effect noted by them can be clearly seen by studying the last two equations in which S_N is increased by both reactions and its increase directly affects both terms in the equation for C_1.

4.8 MICROBIAL LEACHING

4.8.1 Growth of chemolithotrophs on sulphur and metal sulphide ores

Bacteria of the genus. *Thiobacillus* and some others can obtain their energy by the oxidation sulphur and insoluble metal sulphides. The latter process, known as microbial leaching, has been the subject of much interest, as the microbial oxidation of sulphide releases the metal into solution, thus providing an alternative method of extracting metals from their ores. Some minerals contain ferrous iron which the bacteria can oxidise to ferric iron; this in turn can react with the sulphide to release more metal ions (Tuovinen and Kelly, 1974). This purely chemical leaching can occur together with microbial sulphide oxidation. In the chemical process Fe^{3+} is reduced to Fe^{2+} which the bacteria then reoxidise.

Several workers have shown that cells of *T. denitrificans* (Baldensperger *et al,* 1974), *T. thiooxidans* (Schaeffer *et al,* 1963) and *Sulphobus sp.* (Weiss, 1973) growing on pure sulphur are attached to the sulphur particles, and that the sulphur immediately around the cells is considerably eroded. Further it has been demonstrated that attachment to the surface of sulphur crystals is essential for growth of *T. thioxidans* (Schaeffer *et al,* 1963; Beebe and Umbriet, 1971; Cook, 1964). Attachment is also thought to be necessary for growth on metal sulphides (Tuovinen and Kelly, 1974; Gormely *et al,* 1975), the great majority of bacteria being associated with mineral particles during active growth and only appearing in the supernatant at the end of logarithmic growth when most of the substrate has been utilised (McGoran *et al,* 1969; Pinches, 1975). Electron micrographs show that *Thiobacilli* also grow in intimate contact with mineral surfaces and cause local pitting in the surface due to erosion (Tributsch, 1976; Bennett and Tributsch, 1978). Murr and Berry (1976) have shown that with low grade ores bacteria in the genus *Calderella* selectively attach to chalcopyrite and pyrite regions of the ore and not to the siliceous areas of the matrix. They considered attachment to be an an adsorption process. Several authors have demonstrated that *Thiobacilli* produce lipid surfactants which reduce the surface tension of the medium thus wetting the solid surface and allowing bacterial attachment (Beebe and Umbriet, 1971; Jones and Starkey, 1961; Schaeffer and Umbriet, 1963; Cook, 1964). It is likely that similar surfactants are used during growth on metal sulphides, (Duncan, 1967). Some artificial surfactants (e.g. Tween 20) greatly reduce the lag period before growth occurs on metal sulphides, although they have also been shown to inhibit chalcopyrite oxidation (Torma *et al*, 1976). *Thiobacilli* grown on dissolved Fe^{2+} do not produce much (if any) surfactant, which results in a considerable lag before growth occurs after transfer from such a medium to one containing solid sulphide (Duncan, 1967). So it is likely that lipid production is repressed in the absence of solid sulphur/sulphide. Cook (1964) found that *Thiobacilli* required lipids to attach to sulphur particles prior to growth. In the absence of surfactant growth can be greatly inhibited by shaking which apparently restricts cell attachment.

As the organisms are autotrophic, in addition to O_2, they need an adequate supply of CO_2. This can often be a growth limiting factor (Torma *et al*, 1972). Growth and leach rate are directly related (McGoran *et al*, 1969; Pinches, 1975; Gormely *et al*, 1975) and it has been shown repeatedly that growth and leaching are essentially independent of the solids (mineral) concentration (wt/vol) (Pinches *et al*, 1976; Torma *et al*, 1970) but are dependent on the surface area of mineral available (Pinches *et al*, 1976; Pinches, 1975; Gormely *et al*, 1975; Torma *et al*, 1972, Torma *et al*, 1970; Torma and Legault, 1973).

Microbial leaching of ZnS by *T. ferrooxidans* has been modelled by Gormely *et al* (1975) using the hydrocarbon growth model of Erickson *et al* (1970), adopted to continuous culture; leach rates and bacterial growth are proportional to mineral surface area.

The extracellular mechanism used for uptake/oxidation of sulphur and sulphide is as yet unknown (Tuovinen and Kelly, 1974; Bennett and Tributsch, 1978). It is often noted that insoluble materials produced by bacterial action precipitate on the mineral surface thus inhibiting leaching (Silver and Torma, 1974; Tuovinen and Kelly, 1974; Saka guchi *et al*, 1976). Tributsch, 1976; Bennett and Tributsch, 1978).The precipitates are usually sulphates or hydroxides, but can include metals, sulphur and different sulphides. If the mineral is reground to provide fresh surfaces further leaching and growth can occur (Sakaguchi *et al*, 1976). No mathematical models of leaching/growth at present take account of precipitation on the mineral surface as a limiting factor in the process, although a model (Arvin, 1978) exists describing precipitation in a nitrifying film caused by a pH gradient within the film. The mechanism on the mineral surface may be similar.

The overriding importance of the substrate surface and attachment to it is clear, factors affecting it include the source of the inoculum, the presence of surfactants and their production, agitation in the early stages of growth, and the precipitation of insoluble products on the mineral surface.

4.9 SOLID STATE FERMENTATIONS

A solid state fermentation is one in which the substrate is a moist solid, not suspended in water and is often referred to under the Japanese name of a Koji fermentation. In many natural processes such as wood rotting, composting and litter decomposition and food spoilage by moulds, the substrate is also a moist solid not in aqueous suspension. Koji fermentations have been used in the far east for hundreds, and probably thousands, of years for the preparation of certain traditional foods. Some of these processes have been described by Hesseltine and Wang (1967) and also Crank *et al* (1977). In addition Koji fermentations are now being used to produce fungal enzymes such as cellulases, (Toyama, 1976) amylases and proteases. Hesseltine (1967, 1972, 1977a and b) has reviewed some of the work on Koji fermentations including studies on the production of fungal secondary metabolites and SCP from agricultural wastes.

In the traditional processes substrates such as cereal grains, legume seeds or bran are inoculated with spore suspensions and incubated in trays or wrapped in banana leaves or contained in some other utensil. This process has been modernised and improved to give controlled environmental conditions, automatic turning of the fermenting koji and harvesting (Toyama, 1976, Hesseltine, 1972 and 1977). Much of the information on Koji fermentations is in Japanese and is thus not readily available to western scientists (e.g. Terue *et al,* 1959, 1960). Work in the West has concentrated on the production of fungal secondary metabolites, such as aflatoxin and ochratoxin (Hesseltine, 1972, 1977 I), the production of SCP from agricultural, forestry and food industry wastes (Hesseltine 1977 II), Kaneshiro, 1977; Han and Anderson, 1975; Grant *et al,* 1977, 1978; Edwards and Smith, 1975; Smith *et al,* 1975) and the use of fungi to increase protein content of foods such as carob beans (Kokke, 1977) and ground barley (Reade and Smith, 1975). Unfortunately there is little information on rates of fungal growth in Koji fermentations and no systematic work on the kinetics of the process has been found. Typically changes in dry weight, moisture, pH and temperature are measured, together with changes in the relative concentration or amount of various components of the feed stock, which is usually a complex substrate. Thus concentrations of protein, total nitrogen, starch, reducing sugars, cellulose, hemicellulose and lignin may be measured together with the activities of appropriate enzymes such as proteases, amylases and cellulases and the concentration of by-products or waste products like simple sugars, alcohol, amino acids, organic acids and fungal toxins (see for example Hesseltine and Wang, 1967; Hesseltine, 1977 I and II; Crank *et al,* 1977; Wang *et al,* 1974; Kokke, 1977; Edwards and Smith, 1975). Hesseltine (1972, 1977 I and II) stresses the importance of the moisture level which will not only affect performance in pure cultures but will strongly influence the chances of contamination and physical aspects of fermenter operation, e.g. by clogging of substrate. At low moisture bacterial growth tends to be inhibited, thus correct control will minimise the possibility of bacterial contamination. Moisture levels above 12% are required; above this point and below the point at which bacteria are inhibited (less than 35% for Lactic acid bacteria – Hesseltine, 1977, II) moulds will predominant. The rate of agitation has also been shown to be important (Hesseltine, 1977, I) as has temperature (Terui *et al,* 1960). It is likely that the rate of aeration and possibly the oxygen concentration of the air will be important factors as will the substrate size and its pretreatment. In addition to these physical factors the chemical nature of the substrate and the culture environment and pH will obviously be as important as they are to submerged fermentations.

Hesseltine (1972, 1977 II) has listed and discussed the advantages and disadvantages of solid state over submerged fermentations. Toyama and Ogawa (1977) compared cellulase and other enzyme activities produced by several strains of *T. viride*. Solid culture seemed to give better yields although the media used were not identical and much more substrate was available in the solid state culture. Generally, comparisons of yields of biomass, enzymes and metabolites in the two processes are uncommon as are economic comparisons. Recently, however, Pamment *et al* (1978) have compared solid-state (20 to 30% w/v solids) with slurry (1–5% solids) fermentations of alkali-treated sawdust by *Chaetomium cellulolyticum*. Solid-state fermentation produced only slightly less percentage protein content than slurry fermentations, growth (and protein production) rates were however much slower (2–4 times) than in slurries. Despite this the productivity of protein per unit volume of fermentation mixture (g/*l*/hr) in solid state was similar to that in 5% slurry fermentations and two to three times faster than in 1% slurries. Furthermore product recovery costs are lower for solid-state fermentations and it may be that energy costs overall will be cheaper. It is to be hoped that similar quantitative work will soon allow true comparisons to be made between the two methods of fermentation, allowing some conclusions about the economics of the two systems to be drawn.

REFERENCES

Aiba, S., Huang, K. L., Moritz, V. and Someya, J. (1969), *J. Ferment. Technol.*, 47; 211–219.

Akin, D. E. (1976), *Appl. Env. Micro.* **31**; 562–568.

Andreotti, R. E. Mandels, M. and Roche, C. (1977), in *Proc. Bioconversion Symp.* Feb. 1977, Indian Inst. Tech., New Delhi, Ed. Ghose, T. K. 249–268.

Argyle, P. J., Mathison, G. E. and Chandon, R. G. (1976), *J. appl. Bact.* **41**; 175–184.

Arvin, (1978) Internal Report of the Sanitary Engineering Dept. Technical University of Denmark.

Bailey, M., Enari, T. M. and Linko, M. (eds.) (1975) *Symposium on the Enzymatic Hydrolysis of Cellulose,* Aulanko, Finland, March 1975.

Bajpas, R. R., Prokop, A. and Rankrishna, D. (1975), *Biotech and Bioeng.* **17**; 541–556.

Baldensperger, J. L., Guarraia, J. and Humphreys, W. J. (1974) *Arch. Microbiol.* **99**; 322–329.

Beebe, J. L. and Umbriet, W. W. (1971) *J. Bact.* **108**; 612–164.

Beguin, P., Eisen, H. and Roupas, A. (1977), *J. Gen. Microbiol.* **101**; 191–196.

Bellamy, W. D. (1974), *Biotech. and Bioeng.* **16**; 869–880.

Bennett, J. C. and Tributsch, H. (1978), *J. Bact.* **134**; 310–317.

Berg, B., Hofsten, B. V. and Pettersson, G. (1972a) *J. appl. Bact.* **35**; 201–214.

Berg, B., Hofsten, B. V. and Pettersson, G. (1972b) *ibid.*

Berg, B. (1975), *Can. J. Microbiol.* **21**; 51–57.

Berg, B. and Hofsten, A. V. (1976), *J. appl. Bact.* **41**; 395–399.

Berg, B. and Pettersson, G. (1977), *J. appl. Bact.* **42**; 62–75.

Berg, B. (1978), *Arch. Microbiol.* **118**; 61–65.

Blanch, H. W. and Einsele, A. (1973), *Biotech. and Bioeng.* **15**; 861–877.

Breul, C. and Kushner, D. J. (1976), *Can. J. Microbiol.* **21**; 1776–1781.

Brown, D. E., Halstead, D. J. and Howard, P. (1975), p. 137–153 in *Symposium on the Enzymatic Hydrolysis of Cellulose,* Aulanko, Finland, March, 1975, eds. Bailey, M., Enari, T. M. and Linko, M.

Brown, D. E. and Fitzpatrick, S. W. (1976), p. 139–155, in 'Food

from Waste', Proc. 6th Int. Symp. on Food Tech., Weybridge, April 1975, eds Binch, G. G., Parker, K. J. and Worgan, J. (Applied Science Publishers).
Bull, A. T. (1972), *J. appl. Chem. Biotechnol.* **22**; 261–292.
Chahal, D. S., Swain, J. E. and Moo-Young, M. (1977) *Dev. Ind. Microbiol.* **18**; 433–442.
Chakravarty, M., Amin, P., Singh, H. D. Barvah, J. N. and Iyengar, M. S. (1972) *Biotech. and Bioeng.* **14**; 61–73.
Chakravarty, M., Singh, H. D. and Barvah, J. N. (1975) *Biotech. and Bioeng.* **17**; 399–412.
Chang, W. T. H. and Thayer, D. W. (1975), *Dev. Ind. Microbiol.* **16**; 456–464.
Chang, W. T. H. and Thayer, D. W. (1977), *Can. J. Microbiol.* **23**; 1285–1292.
Christensen, P. J. (1977), *Can. J. Microbiol.* **23**; 1599–1653.
Christison, J. and Martin, S. M. (1971), *Can. J. Microbiol.* **17**; 1207–1216.
Coleman, G. (1967), *J. Gen. Microbiol.* **49**; 421–431.
Cook, T. M. (1964), *J. Bact.* **88**; 620–623.
Corbett, N. H. (1965), *J. Inst. of Wood Sci.* **14**; 18–29.
Corpe, W. A. and Winters, H. (1972), *Can. J. Microbiol.* **18**; 1483–1490.
Costerton, J. W., Ingram, J. W. and Cheng, K-J (1974), *Bact. Revs.* **38**; 87–110.
Cowling, E. B. (1975) p. 163–181 in *Biotech. and Bioeng. Symposium No. 5.* ed. Wilke, C. R.
Crank, T. C., Steinkraus, K. H., Hackler, L. R. and Mattick, L. R. (1977) *Appl. Env. Microbiol.* **33**; 1067–1073.
Crawford, D. L. (1974), *Can. J. Microbiol.* **20**; 1069–1072.
Crawford, D. L., McCoy, E., Harkin, J. M. and Jones, P. (1973), *Biotech. and Bioeng.* **15**; 833–843.
Crawford, D. L., Crawford, R. L. and Pometto, A. L. III (1977a) *Appl. Env. Microbiol.* **33**; 1247–1251.
Crawford, D. L., Floyd, S. and Pometto, A. L. III, (1977b), *Can. J. Microbiol.* **23**; 434–440.
Daugulis, A. J. and Bone, D. H. (1977), *European. J. Appl.Micro.* **4**; 159–166.
Day, D. F., Gomersall, M. and Yaphe, W. (1975a), *Can. J. Microbiol.* **21**; 1476–1483.
Day, D. F. and Yaphe, W. (1975b), *Can. J. Microbiol.* **21**; 1512–1518.
Demain, A. L. (1972), 21–32 in *Biotech. and Bioeng. Symposium* No. 3 ed. Wingard, L. B. J.
Dostalek, M. and Molin, N. (1975), p. 385–401 in *Single-Cell Protein II.* eds. Tannenbaum, S. R. and Wang, D. I. C., M. I. T. Press.

Duckworth, M. and Turvey, J. R. (1969), *Biochem. J.* **113**; 139–142.
Duncan, D. W. (1967), *Aust. Min.* **59**; 21–25.
Dunlap, C. E. (1975), p. 244–262 in *Single-Cell Protein II.* eds. Tannenbaum, S. R. and Wang, D. I. C., M. I. T. Press.
Dunlap, C. E. Thomson, J. and Chang, L. C. (1976) *A.I.Ch.E. Symposium Series.* 158; Vol. **72**; 58–63.
Dworkin, M. (1966), *Ann. Revs. Microbiol.* **20**; 75–106.
Edwards, J. R. and Smith, R. N. (1975), p. 545–551, *3rd International Biodegradation Symposium.* eds. Sharpley, J. M. and Kaplan, A. M., Appl. Sci. Publishers.
Einsele, A. and Fiechter, A. (1975a), p. 169–194 in *Adv. in Biochemical Engineering.* **1**; eds. Ghose, T. K. and Fiechter, A , Springer-Verlag.
Einsele, A., Schneider, H. and Fiechter, A. (1975b) *J. Ferment. Technol.* **53**; 241–243.
Ek, M. and Eriksson, K.-E. (1975), *Appl. Polymer Symposium.* **28**; 197 –203.
Enari, T-M., Markannen, P. and Korhonen, E. (1975), p. 171–180 in *Symposium on the Enzymatic Hydrolysis of Cellulose.* Aulanko, Finland, March 1975 eds. Bailey, M., Enari, T-M. and Linko, M.
Enebo, L. (1949), *Nature.* **163**; 805.
Erickson, L. E. and Humphrey, A. E. (1969a) *Biotech. and Bioeng.* **11**; 449–466.
Erickson, L. E. and Humphrey, A. E. (1969b) *Biotech. and Bioeng.* **11**; 467–487.
Erickson, L. E. and Humphrey, A. E. (1969c) *Biotech. and Bioeng.* **11**; 489–515.
Erickson, L. E., Fan, L. T., Shah, P. S. and Chen, M. S. K., (1970) *Biotech. and Bioeng.* **12**; 713–746.
Erickson, L. E. and Nakahara, T. (1975a), *Process Biochem.* **10**; Part, 5, (June), 9–13.
Eriksson, K-E, and Larsson, K. (1975b), *Biotech. and Bioeng.* **17**; 327–348.
Eriksson, K-E (1978), *Biotech. and Bioeng.* **20**; 317–332.
Forage, A. J. and Righelato, R. C. (1978) p. 59–94. *Progress in Industrial Microbiology.* **14**; ed. Bull, M. J., Elsevier Scientific Publishing Co.
Fu, T. T. and Thayer, D. W. (1975), *Biotech. and Bioeng.* **17**; 1749–1760.
Fusee, M. C. and Leatherwood, J. M. (1972), *Can. J. Microbiol.* **18**; 347–353.
Gaden, E. L. Jr., Mandels, M., Reese, E. T. and Spano, L. A. (eds.) (1976) 'Enzymatic Conversion of Cellulosic Materials: Technology

and Applications', *Biotech. and Bioeng. Symposium No. 6.*

Ghose, T. K., Pathak, A. N. and Bisaria, V. S. (1975) p. 111–136, in Symposium on the Enzymatic Hydrolysis of Cellulose, Aulanka, Finland, March, 1975, eds. Bailey, M., Enari, T-M, and Linko, M.

Ghose, T. K. (ed.) (1977) *Proc. Bioconversion Symposium,* Indian Inst. Tech., New Delhi, Feb. 1977.

Goma, G., Pareilleux, A. and Durand, G. (1973) *J. Ferment. Technol.* **51**; 616–618.

Gormely, L. S., Duncan, D. W., Branion, R. M. S. and Pinder, K. L. (1975) *Biotech. and Bioeng.* **17**; 31–49.

Grant, G. A., Anderson, A. W. and Han, Y. W. (1978) *Biotech. and Bioeng.* **19**; 1817–1830.

Grant, G. A., Han, Y. W. and Anderson, A. W. (1978) *Appl. Env. Microbiol.* **35**; 549–553.

Greaves, H. (1971) *Aust. J. Biol. Sci.* **24**; 1169–1180.

Groleau, D. and Yaphe, W. (1977), *Can. J. Microbiol.* **23**; 672–679.

Gupta, D. P. and Heale, J. B. (1971), *J. Gen. Microbiol.* **63**; 163–173.

Gutierrez, J. R. and Erickson, L. E. (1977), *Biotech. and Bioeng.* **19**; 1331–1349.

Hackett, W. F., Connors, W. J., Kirk, T. K. and Zeikus, J. G. (1977) *Appl. Env. Microbiol.* **33**; 43–51.

Hajny, G. J. and Reese, E. T. (eds.((1969), 'Cellulases and their Applications' *Adv. in Chem. Series No. 95.* (ACS).

Halliwell, G. and Griffin, M. (1973), *Biochem. J.* **135**; 587–594.

Han, Y. W. and Callihan, C. D. (1974), *Appl. Micro.* **27**; 159–165.

Han, Y. W. and Anderson, A. W. (1975), *Appl. Micro.* **30**; 930–934.

Hancock, J. G., Eldridge, C. and Alexander, M. (1970), *Can. J. Microbiol.* **16**; 69–74.

Hesseltine, C. W. and Wang, H. L. (1967) *Biotech. and Bioeng.* **9**; 275–288.

Hesseltine, C. W. (1972), *Biotech. and Bioeng.* **14**; 517–532.

Hesseltine, C. W. (1977a), *Process Biochem.* July/August, 24–27.

Hesseltine, C. W. (1977b), *Process Biochem.* Nov. 29–32.

Highley, T. L. (1973), *Wood and Fibre.* **5**; 50–58.

Highley, T. L. (1975), *Forest Prod. J.* **25**; (7), 38–39.

Highley, T. L. (1976), *Material und Organismen.* **11**; 33–46.

Highley, T. L. (1977), *Material und Organismen.* **12**; 25–36.

Hodgkiss, W. and Shewan, J. M. (1968), p. 127–166 in *Advances in the Microbiology of the Sea.* ed. Droop, M. R. and Ferguson-Wood, E. J., Academic Press.

Hofsten, B. V., Berg, B. and Beskow, S. (1971), *Arch. Microbiol.* **79**; 69–79.

Hofsten, B. V. and Malmquist, M. (1974), *J. Gen. Microbiol.* **87**; 150–158.

Hofsten, B. V. (1975), p. 281–297, in *Symposium on the Enzymatic Hydrolysis of Cellulose.* Aulanko, Finland, March 1975. Eds. Bailey, M., Enari, T-M and Linko, M.

Horvath, R. S. (1972), *Bact. Revs.* **36**; 146–155.

Howell, J. A. and Mangat, M. (1978), *Biotech. and Bioeng.* **20**; 847–863.

Howell, J. A. and Stuck, J. (1975), *Biotech. and Bioeng.* **17**; 873–893.

Huang, A. A. (1975), *Biotech. and Bioeng.* **17**; 1421–1433.

Hug, H., Blanch, H. W. and Fiechter, A. (1974), *Biotech. and Bioeng.* **16**; 965–985.

Hulme, M. A. and Stranks, D. W. (1971), *J. Gen. Microbiol.* **69**; 145–155.

Humphrey, A. E. (1975), p. 1–23 in *Single-Cell Protein II.* eds. Tannenbaum, S. R. and Wang, D. I. C., M.I.T. Press.

Humphrey, A. E., Moreira, A., Armiger, W. and Zabriskie, D. (1977) p. 45–64, in *Biotech. and Bioeng. Symposium No. 7.* eds. Gaden, E. L. Jr. and Humphrey, A. E.

Hungate, R. E. (1947) *J. Bact.* **53**; 631–645.

Hungate, R. E. (9166) *The Rumen and its Microbes.* Academic Press N. Y. and London.

Jane, F. W. (1970) *The Structure of Wood.* Adam and Charles Black, London.

Jones, G. E. and Starkey, R. L. (1961) *J. Bact.* **82**; 788–789.

Jurasek, L., Colvin, J. R. and Whitaker, D. R. (1967) p. 131–170. in *Adv. in Appl. Microbiol.* **9**; ed. Umbriet, W. W.

Kaeppeli, O. and Fiechter, A. (1976) *Biotech. and Bioeng.* **18**; 967–974.

Kaneshiro, T. (1977), *Dev. Ind. Microbiol.* **18**; 591–597.

Keay, L., Moseley, M. H., Anderson, R. G., O'Connor, R. J. and Wildi, B. S., (1972) p. 63–92, *Biotech. and Bioeng. Symposium No. 3.* ed. Wingard, L. B. Jr.

Keilich, G., Bailey, P. & Liese, W. (1970) *Wood Sci. & Technol.* **4**; 273–283.

Kennedy, R. S., Finnerty, W. R., Sudarsanank and Young, R. H., (1975), *Arch. Microbiol.* **102**; 75–83.

Khan, A. W., (1977), *Can. J. Microbiol.* **23**; 1700–1705.

Kim, C. (1974), presented at the Chemical Reaction Engineering Conference, Munich, September 1974.

King, K. W. (1959), *J. Dairy Sci.* **42**; 1848–1856.

Kirk, T. K. (1971) *Ann. Revs. Phytopath.* **9**; 185–210.

Kirk, T. K. (1975), *Biotech. and Bioeng. Symposium.* No.5, 139–150.

Kirk, T. K., Connors, W. J. and Zeikus, J. G. (1976) *Appl. Env. Microbiol.* **32**; 192–194.

Kirk, T. K., Schaltz, E., Connors, W. J., Lorenz, L. F. & Zeikus, J. G. (1978) *Arch. Microbiol.* **117**; 277–285.

Kokke, R. (1977) *J. appl. Bact.* **43**; 303–307.

Krishnamurti, G. R. and Kitts, W. D. (1969), *Can. J. Microbiol.* **15**; 1373–1379.

Kurowski, W. M. and Dunleary, J. A. (1976) *J. appl. Bact.* **41**; 119–128.

Lampen, J. O. (1965), p. 115–133 in *'Function and Structure in Microorganisms', 15th Symposium of the Soc. Gen. Microbiol.* eds Pollock, M. R. and Richmond, M. H., C.U.P.

Lampen, J. O. (1978), p. 231–247 in 'Relations between Structure and Function in the Prokaryotic Cell', *28th Symposium of the Soc. Gen. Microbiol.* eds. Stanier, R. Y., Rogers, H. J. and Ward, J. W., C.U.P.

Latham, M. J., Brooker, B. E., Pettifer, G. L. and Harris, P. J. (1978a), *Appl. Env. Microbiol.* **35**; 156–165.

Latham, M. J., Brooker, B. E., Pettifer, G. L. and Harris, P. J. (1978b), *Appl. Env. Micro.* **35**; 1166–1173.

Lee, S. E., Armiger, W. B., Watteeuw, C. M. and Humphrey, A. E. (1978) *Biotech. and Bioeng.* **20**; 141–144.

Levisohn, S. and Aronson, A. I. (1967), *J. Bact.* **93**; 1023–1030.

Liese, W. (1970), *Ann. Revs. Phytopath.* **8**; 231–258.

Lilley, G. & Bull, A. T. (1974), *J. Gen. Microbiol.* **83**; 123–133.

Lisker, N., Katan, J., Chet, I. and Henis, Y. (1975), *Can. J. Microbiol.* **21**; 521–526.

McGoran, C. J. M., Duncan, D. W. and Walden, C. C. (1969) *Can. J. Microbiol.* **15**; 135–138.

Mallee, F. M. and Blanch, H. W. (1977), *Biotech. and Bioeng.* **19**; 1793–1816.

Mandels, M. and Weber, J. (1969), p. 391–413 in 'Cellulases and their Applications', *Adv. Chem. Series No.95.* eds. Hajney, G. J. and Reese, E. T. (A.C.S.).

Mandels, M., Weber, J. and Parizek, R. (1971a), *Appl. Micro.* **21**; 152–154.

Mandels, M., Kostick, J. and Parizek, R. (1971b), *J. Polymer. Sci. part C.* **36**; 445–459.

Mandels, M., Hontz, L. and Nystrom, J. (1974), *Biotech. and Bioeng.* **16**; 1471–1493.

Mandels, M. (1975), p. 81–105, in *Biotech. and Bioeng. Symp. No. 5.* ed. Wilke, C.

Mangat, N. and Howell, J. A. (1978), *A.I.Ch.E. Symposium Series.* No. 172, Vol. **74**; 77–81.

Marshall, J. J. (1973), *Carbohydrate Research.* **26**; 274–277.
Martin, S. M. and So, V. (1969), *Can. J. Microbiol.* **15**; 1393–1397.
Maxham, J. V. and Maier, W. J. (1978a) *Biotech. and Bioeng.* **20**; 865–898.
Maxham, J. V. and Maier, W. J. (1978b) *Biotech. and Bioeng.* **20**; 1745–1744.
Meylan, B. A. and Butterfield, B. G. (1972) *Three-Dimensional Structure of Wood – A Scanning Electron Microscope Study.* Chapman & Hall, London.
Millet, M. A., Baker, A. J. and Salter, L. D. (1975) p. 193–219 in *Biotech. and Bioeng. Symposium. No.5.* ed. Wilke, C. R.
Mimura, A., Watanabe, S. and Takedu, J. (1971) *J. Ferment. Technol.* **49**; 225–271.
Miura, Y., Okazaki, M., Hamada, S-I, Murakawa, S-I, and Yugan, R. (1977a) *Biotech. and Bioeng.* **19**; 701–714.
Miura, Y., Okazaki, M., Murakawa, S-I, Hamada, S-I, and Ohro, K. (1977b) *Biotech. and Bioeng.* **19**; 715–720.
Monreal, J. and Reese, E. T. (1969), *Can. J. Microbiol.* **15**; 689–696.
Montenecourt, B. S. and Eveleigh, D. E. (1977a), *Appl. Env. Micro.* **33**; 178–183.
Montenecourt, B. S. & Eveleigh, D. E. (1977b), *Appl. Env. Micro.* **34**; 777–782.
Moo-Young, M., Shimizu, T. and Whitworth, D. A. (1971a) *Biotech. and Bioeng.* **13**; 741–760.
Moo-Young, M., & Shimizu, T. (1971b) *Biotech. and Bioeng.* **13**; 761–778.
Moo-Young, M. (1976), *Process Biochem.* **11**; Dec. 1976, 32–34.
Moo-Young, M., Chahal, D. S., Swain, J. E. and Robinson, C. W. (1977a) *Biotech. and Bioeng.* **19**; 527–538.
Moo-Young, M. (1977b), *Process Biochem.* **12**; No.4, May 1977, 6–10.
Moo-Young, M., Chahal, D. S. and Vlach, D. (1978), *Biotech. and Bioeng.* **20**; 107–118.
Murr, L. E. and Berry, V. K. (1976), *Hydrometallurgy.* **2**; 11–24.
Nakahara, T., Erickson, L. E. and Gutierrez, J. R. (1977), *Biotech. and Bioeng.* **19**; 9–25.
Nesse, N., Wallick, J. and Harper, J. M. (1977), *Biotech. and Bioeng.* **19**; 323–326.
Nilsson, T. (1973), *Studia Forestalia Suecica, No.104.*
Nilsson, T. (1974), *Material und Organismen.* **9**; 173–198.
Nilsson, T. (1975) p. 103–112 in 'Organismen und Holz', eds. Becker, G. and Liese, W. Suppliment (Heft 3) to *Material und Organismen.*
Nisizawa, T., Suzuki, H. and Nisizawa, K. (1972), *J. Biochem.* **71**; 999–1007.

Nolan, E. J. and Forro, J. R. (1976), p. 432 in *Abst. of Papers given at 5th Int. Fermentation Symp.*, Berlin, 1976, ed. Dellweg, M.

Norberg, P. and Hofsten, B. V. (1969), *J. Gen. Microbiol.* **55**; 251–266.

Norkrans, B. (1967), p. 91–130 in *Adv. in Appl. Micro.* **9**; ed. Umbriet, W.W.

Noval, J. J. and Nickerson, W. J. (1959), *J. Bact.* **77**; 251–263.

Ogelsby, R. T., Christman, R. F. and Driver, C. H. (1967), *Adv. in Appl. Microbiology.* **9**; 171–184. ed. Umbriet, W. W. Academic Press.

Okazaki, M. and Moo-Young, M. (1978) *Biotech. and Bioeng.* **20**; 637–663

Okazaki, M., Miyamato, Y. and Moo-Young (1979a), to be published.

Okazaki, M. and Moo-Young, M. (1979b), to be published.

Ornston, L. N. (1971), *Bact. Revs.* **35**; 87–116.

Osmunsvag, K. and Goksoyr, J. (1975), *Eur. J. Biochem.* **57**; 405–409.

Osumi, M., Fukazumi, F., Ternaishi, Y., Tanaka, A., and Fukui, S, (1975) *Arch. Microbiol.,* **103**; 1–11.

Pamment, N., Robinson, C. W., Hilton, J. and Moo-Young, M. (1978) *Biotech. and Bioeng.* **20**; 1735–1744 (1978).

Patterson, H., Irvin, R., Costerton, J. W. & Cheng, K. J. (1975), *J. Bact.* **122**; 278–287.

Paredes-Lopez, O. and Gonzalez, Y. (1973), *J. Ferment. Technol.* **51**; 619–623.

Peitersen, N. (1975a), *Biotech. and Bioeng.* **17**; 361–374.

Peitersen, N. (1975b), *Biotech. and Bioeng.* **17**; 1291–1299.

Peitersen, N. (1977), *Biotech. and Bioeng.* **19**; 337–348.

Pinches, A. (1975) p. 28–35 in *Leaching and Reduction in Hydrometallurgy.* ed. Burkin, A. R. (IMM symposium, London 1975).

Pinches, A., Al-Jaid, F. O., Williams, D. J. A. and Atkinson, B. (1976) *Hydrometallurgy.* **2**; 87–103.

Polcin, J. & Bezuch, B. (1977) *Wood Sci. and Technol.* **11**; 275–290.

Pollock, M. R. (1962) Chapter iv p. 121–178, in *The Bacteria - Vol. IV: The Physiology of Growth.* eds. Gunwalus, I. C. and Stanier, R. Y., Academic Press.

Prokup, A., Erickson, L. E. and Paredes-Lopez, O. (1971) *Biotech. and Bioeng.* **13**; 241–256.

Rapp, P. and Wagner, F. (1976), p. 133 in *Abst. 5th International Fermentation Symp.* Berlin, 1976, ed. Dellweg, H.

Reade, A. E. and Smith, R. H. (1975) *J. appl Chem. and Biotech.* **23**; 785–799.

Reese, E. T. (1972), p. 43–62, *Biotech. and Bioeng. Symposium No. 3.* ed. Wingard, L. B., Jr.

Reese, E. T., Maguire, A. and Parrish, F. W. (1973), *Can. J. Microbiol.* **19**; 1065–1074.

Rogers, C. J., Coleman, E., Spiro, D. F. and Purcell, T. C. (1972) *Env. Sci. and Tech.* **6**; 715–719.

Rogers, H. J. (1970), *Bact. Revs.* **34**; 194–214.

Rogers, H. J., Barrie Ward, J. and Burdett, I. D. I. (1978), p. 139–176, in 'Relations between Structure and Function in the Prokaryotic Cell', *28th Symposium of the Soc. Gen. Microbiol.* eds. Stanier, R. Y., Rogers, H. J. and Ward, J. B.

Romantschuk, H. (1975), p. 344–356 in *Single-Cell Protein II.* eds. Tannenbaum, S. R. and Wang, D. I. C., M.I. T. Press.

Rowlands, S. P. (1975) p. 183–191 in *Biotech. and Bioeng. Symposium No.5.* ed. Wilke, C. R.

Rossell, S. F., Abbot, E. G. M. and Levy, J. I. (1973) *J. Inst. Wood Sci.* No.32 (Vol.6 No.2) 28–35.

Sakaguchi, H., Silver, M. and Torma, A. E. (1976), *Biotech. and Bioeng.* **18**; 1091–1101.

Santos, T., Villanveva, J. R. and Nombela, C. (1977), *J. Bact.* **129**; 52–58.

Schaeffer, W. I. and Holbert, P. E. and Umbriet, W. W. (1963a), *J. Bact.* **85**; 137–40.

Schaeffer, W. I. and Umbriet, W. W. (1963b), *J. Bact.* **85**; 492–493.

Seifter, S. and Harper, E. (1970), p. 613–635, *Methods in Enzymology.* **19**; *Proteolytic Enzymes.* ed. Perlman, G. E. and Lorand, L., Academic Press, N. Y. and London.

Sheehan, B. T. and Johnson, M. J. (1971), *Appl. Micro.* **21**; 511–515.

Shilo, M. (1970), *J. Bact.* **104**; 453–461.

Silver, M. and Torma, A. E. (1974), *Can. J. Microbiol.* 141–147.

Skinner, F. A. (1972), *J. Appl. Bact.* **35**; 453–462.

Smith, R. E. and Nendoerffer, T. S. (1971), *Can. J. Microbiol.* **17**; 31–37.

Smith, R. N., Odell, D. E. and Edwards, J. R. (1975), p. 553–559 in *3rd International Biodegradation Symposium.* eds. Sharpley, J. M. and Kaplan, A. M., Appl. Sci. Publishers.

Smith, W. R., Yu I. and Hungate, R. E. (1973), *J. Bact.* **114**; 729–737.

Srinivasan, V. R. (1975), p. 393–405 in *Symposium on the Enzymatic Hydrolysis of Cellulose.* Aulanko, Finland, March 1975, eds. Bailey, M., Enari, T. M. and Linko, M.

Stamatoudis, M. and Tavlarides, L. L. (1973), Paper 24C, 75th National Meeting A. I. Ch. E., Detroit.

Stanier, R. Y. (1942), *Bact. Revs.* **6**; 143–196.

Starr, M. P. and Seider, R. J. (1971), *Ann. Revs. Microbiol.* **25**; 649–675.

Sternberg, D. V., Jayakumar, D. and Reese, E. T. (1977), *Can. J. Microbiol.* **23**; 139–147.

Stewart, B. J. and Leatherwood, J. M. (1976), *J. Bact.* **128**; 609–615.

Stolp, H. and Starr, M. P. (1965), *Ann. Revs. Microbiol.* **19**; 79–104.

Stone, J. E. and Scallan, A. M. (1968), *Cellulose Chem. Technol.* **2**; 343–358.

Stone, J. E., Scallan, A. M., Donefer, E. and Ahlgren, E. (1969a), p. 219–241, in 'Cellulases and their Applications', *Adv. Chem. Series.* **95**; ed. Hajny, G. J. and Reese, E. T. (A.C.S.).

Stone, J. E., Treiber, E. and Abrahamson, B. (1969b), *Tappi.* **52**; 108–110.

Suga, K., van Dedem, G. and Moo-Young, M. (1975), *Biotech. and Bioeng.* **17**; 185–20.

Sundarraj, N. and Bhat. J. V. (1972), *Arch. Microbiol.* **85**; 159–167.

Su, T-M. and Paulavicius, I. (1975) A.I.Ch.F. *Symposium Series.* No. 158, **72**; 72–76.

Sutherland, I. W. (1972), *Adv. Microbiol. Physiol.* **8**; p. 143–231, eds Rose, A. H. and Tempest, D. W., Academic Press.

Suzuki, H. (1975), p. 155–169, in *Symposium on the Enzymatic Hydrolysis of Cullulose.* Aulanko, Finland, March 1975, eds. Bailey, M., Enari, T-M, and Linko, M.

Suzuki, T., Tanaka, K., Matsubara, I. and Kinoshita, S. (1969), *Agr. Biol. Chem.* **33**; 1619–1627.

Stuck, J. D. and Howell, J. A. (1973) Water, 1973, *CEP Symp. Series* **70**; 136, 337–349.

Tannenbaum, S. R. and Wang, D. I. C. (1975), eds. *Single-Cell Protein II.* M.I.T. Press.

Tarkow, H. and Feist, W. L. (1969), 197–218 in 'Cellulases and their Applications', *Adv. Chem. Series.* **95**; (ACS) ed. Hajney, G. J. and Reese, E. T.

Terui, G., Skibaski, I. and Mochizuki, T. (1959), *J. Ferment. Technol* **37**; 479–483.

Terui, G., Shibaski, I. and Mochizuki, T. (1960), *J. Ferment. Technol.* **38**; 29–39.

Terui, G. (1972), p. 33–36 in *Biotech. and Bioeng. Symposium No. 3.* ed. Wingard, L. B. Jr.

Thayer, D. W. (1976), *Dev. Ind. Microbiol.* **17**; 79–89.

Thayer, D. W. (1978), *J. Gen. Microbiol.* **106**; 13–18.

Torma, A. E., Walden, L. L. and Branion, (1970), *Biotech. and Bioeng.* **12**; 501–517.

Torma, A. E., Walden, L. L., Duncan, D. W. and Branion, R. M. R. (1972), *Biotech. and Bioeng.* **14**; 777–786.

Torma, A. E. and Legault, G. (1973), *Ann. Microbiol. (Inst. Pasteur)* **124A**; 111–121.

Torma, A. E., Gabra, G. G., Guay, R. and Silver, M. (1976), *Hydrometallurgy.* **1**; 301–309.

Toyama, N. (1976), p. 207–219 in *Biotech. and Bioeng. Symposium No. 6.* eds. Gaden, E. *et al.*

Toyama, N. and Ogawa, K. (1977), p. 305–327 in *Proc. Bioconversion Symp.* Feb. 1977, Indian Inst. Tech. New Delhi, ed. Ghose, T. K.

Tributsch, H. (1976), *Naturwissenschaften.* **63**; 88.

Tuovinen, O. H. and Kelly, D. P. (1974), *International Metallugical Reviews.* **19**; 21–31.

Turvey, J. R. and Christison, J. (1967), *Biochem. J.* **105**; 311–316.

Updegraff, D. M. (1971), *Biotech. and Bioeng.* **13**; 77–97.

Van Dyke, B. H. Jr. (1972) *Enzymatic Hydrolysis of Cellulose - A Kinetic Study.* PhD. thesis, M.I.T.

Velankar, S. H., Barnett, S. M., Houston, C. W. and Thompson, A. R. (1975), *Biotech. and Bioeng.* **17**; 241–251.

Wang, D. I. C. and Ochoa, A. (1972), *Biotech. and Bioeng.* **14**; 345–360.

Wang, H. L. (1967), *J. Bact.* **93**; 1794–1799.

Wang, H. L., Vespa, J. B. and Hesseltine, L. W. (1974), *Appl. Micro.* **27**; 906–911.

Watson, J. D. (1977), *Molecular Biology of the Gene.* 3rd ed. W. A. Benjamin Inc.

Weimer, P. J. and Zeikus, J. G., (1977), *Appl. Env. Micro.* **33**; 289–297.

Weiss, R. L. (1973), *J. Gen. Microbiol.* **77**; 501–507.

Wheatley, M. A. and Moo-Young, M. (1977) *Biotechnol. and Bioeng.* **19**; 219–233.

Whitworth, D. A., Moo-Young, M. and Viswanatha, T. (1973), *Biotech. and Bioeng.* **15**; 649–675.

Wilcox, W. W. (1968), U.S. Forest Research Paper, FPL 70, July 1968.

Wilke, C. R. (ed.) (1975), 'Cellulose as a Chemical and Energy Resource', *Biotech. and Bioeng. Symposium No. 5.*

Wodzinski, R. S. and Bertolini, D. (1972), *Appl. Micro.* **23**; 1077–1081.

Wodzinski, R. S. and Coyle, J. E. (1974), *Appl. Micro.* **27**; 1081–1084.

Wohner, G. and Wober, G. (1978), *Arch. Microbiol.* **116**; 303–310.

Wood, T. M. (1975), p. 111–137 in *Biotech. and Bioeng. Symposium No. 5.* ed. Wilke, C. R.

Yamane, K., Suzuki, H. and Nisizawa, K. (1970), *J. Biochem.* **67**; 19–35.

Yamane, K., Yoshikawa, T., Suzuki, H. and Nisizawa, K. (1971), *J. Biochem.* **69**; 771–780.

Yaphe (1975) ?

Yoshida, F., Yamane, T. and Yagi, H. (1971a), *Biotech. and Bioeng.* **13**; 215–228.

Yoshida, F. and Yamane, T. (1971b), *Biotech. and Bioeng.* **13**; 691–695.

Yoshida, F., Yamane, T. and Nakamoto, K-I, (1973), *Biotech. and Bioeng.* **15**; 257–270.

Yoshida, F. and Yamane, T. (1974), *Biotech. and Bioeng.* **16**; 635–657.

Zajic, J. E., Guignard, H., and Gerson, D. F. (1977), *Biotech. and Bioeng.* **19**; 1285–1301.

Chapter 5

The Measurement of Process Variables

Dr. Alan Vincent, Dymor, Sherborne, Dorset, and
Dr. Gordon Priestley, Gillingham, Dorset.

5.1 INTRODUCTION

In this review of the 'state of the art' we have assumed that the area of interest is that of fermentation as generally understood: an aerobic (now rarely anaerobic) process for the production of living cells or their products in suspended culture.

'Process variables' are thus those affecting this area, and we have attempted to cover the more important ones. Measurement systems developed purely for *study* of cell biology have not been included, except some in which the authors consider could be translated from the laboratory bench to form part of a continuous monitoring system.

We hope that this summary of fermentation process measurement will serve as a useful introduction to the subject for those meeting its range of problems for the first time, whether they be biologists, microbiologists, chemists or engineers by training. For those already involved in practical applications of these techniques, we trust that the summary offered here will be refreshing and useful, and that at least some new perspectives will be opened.

One basic problem is universal in applying the techniques here described to actual production processes: most of the variables of interest are not themselves electrical or pneumatic quantities. The conversion of the variables to such analogue forms may involve certain assumptions about the performance of the sensors or signal processing equipment: these assumptions should never be forgotten when basing decisions or generalisations on data obtained using such equipment.

Another difficulty may be introduced by the fact that not all measurements can be made on a 'real-time' basis suitable for use in direct feedback or feed-forward control. Chemical analyses are one such group, and where these consume some time before results are available, a closed-loop feedback control system would have the effector part of the loop reacting always to correct an *historic* error; stable operation is therefore impossible. A similar effect is produced in correcting action produces only delayed response in the process due to, for example, slow metabolic response or long mixing time. Relatively simple modification to control loops by introducing a delay, chosen empirically, improves stability. Some form of interactive control using computer, microprocessor or even hard-wired signal processing may be necessary when reaction times vary enormously, for example in batch growth where substrate uptake, acid formation, and indeed all metabolism-dependent changes will occur very much faster (up to 10^4 or more times) at peak population than just after inoculation of a batch culture. Interactive systems based on digital data processing may alter the control loop reactivity following stored experience of system reaction; simpler interactive systems may simply have wired-in programming to alter reactions as a function of time or of some observed value, e.g. turbidity.

Cases exist (Vincent, personal observations) where samples are automatically withdrawn from fermenters and analysed continuously (e.g. sugar) or intermittently (e.g. protein) and automatic control is effected intermittently with uncontrolled periods in between. True interactive control would adjust the amount of control action taken in accordance with the size of error, the rate of change and experience of the process in general: i.e. using many of the refinements normally applied by a human operator using direct manual control!

The general class of system mentioned above, based on *withdrawal* of samples, avoids many of the physical problems (robustness, sterility, calibration drift) associated with direct sensing, and it is often possible to use one set of complicated and expensive monitoring equipment shared over several production vessels. The problems of long time constants thus actually *permit* such time sharing without any loss of fineness of control. Normally variables such as temperature, airflow, and motor power would be controlled locally in closed-loop systems for each fermenter, since for 'simple' variables this is actually cheaper than by multiplexing one set of control gear over all fermenters (but see MacLennan, 1970).

Partial or complete 'automation' of a fermentation (or of any other) process may confer large quantitative and often qualitative benefits for all involved. Organisational problems may originate, however, in the hostility that the change may provoke if it threatens job security; this is a management problem that must be solved for each particular case. A second cause of hostility which is at least as serious, existing as it does also in new installations, is that the use of the equipment may reduce staff involvement, removing decision-making to higher levels in the organisation. Simple examples are the amount of soda, the flow of cooling water etc. to be used in particular situations and being used at any one moment. Such knowledge of what to do and what is happening were once the jealously guarded fruits of often painful experience.

In an automated plant the personal experience is not so obviously necessary, but it should be exhaustively drawn upon in setting up not only the mechanical side of the control system but also the schedule for the operation of the plant. It should be assured that personnel at all reasonable levels will be in touch with what is happening in the plant.

A strong recommendation must be made that the data from the measurement and analysis should be available in the plant itself; the practical and psychological value of this is enormous. It also allows better decentralisation of decision-making, avoiding such nonsense as international telephone calls to an executive to ask if a fermenter should be harvested! It is in theory possible to convert all data aquired in the plant to electrical signals which are transmitted in analogue or (preferably) digital form to an administration block where the data processing and perhaps some of the feedback control reactions are initiated. Such a system is unacceptable because of the separation of functions which it creates or exacerbates, and also because of poor reactions to unforeseen conditions.

The ideal layout is one in which the laboratory/control room is part of the plant, and analysis, data processing and all control functions, with as far as possible also production planning, are performed on the spot. For practical reasons microbiological work may have to be performed elsewhere, but this is to be regretted. 'Teach-ins' about the running of the plant should be held in working hours to keep all personnel up to date with effective use of the system as a whole. Such a measure may be seen by some to be a security risk, but the advantages far outweigh such negative considerations.

The following sections should be read bearing in mind the above remarks on implementation.

5.2 FACTORS AFFECTING FERMENTATION PROCESSES

5.2.1 Substrates

According to the work of Monod (1950) and of Herbert *et al* (1956) the specific growth rate of an organism is controlled, at each level of growth medium concentration and composition, by the concentration of a single limiting nutrient. This statement assumes viability, a suitable physical environment for growth and division, absence of predation, and no inhibition by end products or other toxic materials. Within the concentration range in which it is limiting, a nutrient may be expected to influence growth rate in the following manner:

$$\mu = \mu_{max} \cdot \frac{S}{S + K_s}$$

where μ is the growth rate, μ_{max} is the maximum growth rate, S is the substrate concentration and K_s=S when $\mu = ½ \mu_{max}$, from which it is seen that the relationship is by no means linear across the full range of S values. In the normal operating range (commercial practice) where $S \gg K_s$, the relationship approximates to linearity.

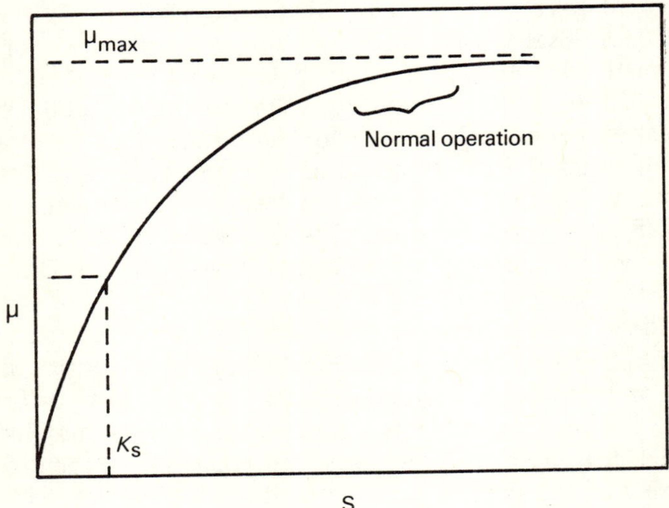

Fig. 5.1
Relationship of growth rate to concentration of limiting substrate.

S values much higher than K_s are used to ensure maximum growth rate and to obviate the need for continuous feeding of the culture. There are instances, however, where limitation of substrate concentration achieves better conversion to cell material or to other product (e.g. Hospodka (1966)).

The similarity of the Monod rate equation to the classical enzyme rate equation of Michaelis and Menten (1913) is assumed to result from the existence of a single growth-rate-controlling step, such as

might be concerned with transport or metabolism of the nutrient concerned, and there is indeed evidence for the existence of transport enzymes in the cell wall. For a detailed discussion of 'Monod' kinetics and of subsequent attempts to amplify this theory and its mathematical treatment to include maintenance energy, multiple substrates, oxygen, cell viability, inhibition, wall growth, etc., see Barford and Hall (1978).

The set of variables comprising the concentrations of the various nutrients must at least be recognised in any serious attempt to optimise a process in terms of yield. In most cases the situation may be simplified by using a non-inhibitory excess of many nutrients and paying close attention to only those with a decisive role either in the metabolism or otherwise in process profitability.

It is evident that in a batch process conditions which were optimal at start-up may no longer be so later in the growth cycle, and that either an excess must be initially present or supplementary feed must be provided during the run.

The main nutrients to be considered later are sources of carbon and of nitrogen. Those nutrients to be continuously added (oxygen, methane, sometimes glucose) or which may be inhibitory (methanol, phenol, many industrial wastes) need special consideration. As with many biological systems, the theoretical consideration of a steady state is rather simpler than for an actual system changing with time (Pirt, 1969), but is rarely simpler in practice. The special case of oxygen as a nutrient is exhaustively covered by Wimpenny (1969) and Nagai *et al* (1971), control of oxygen supply by Vincent (1974a), and the interrelation of rheology/mixing with oxygen supply by Vincent & Priestley (1975).

5.2.2 Product effects

The effects of product concentration on rates of growth and of further product formation are rather more varied in type than are the effects of nutrients. Many simple products may be directly inhibitory (Peringer, 1972), for example ethyl alcohol and lactic acid, but may be re-metabolised at a later stage of a batch fermentation. In some cases the product-forming reactions may be reversible (Humphrey, 1974), and a slowing down of net product formation is observed as would be expected from classical mass action kinetics. Another type of product effect is inhibition caused by physical change, for example of pH value. A large improvement in yield is seen in buffered or pH-controlled cultures when compared to an uncontrolled system (e.g. Moraine and Rogovin, 1971). Products of growth may also affect rheology, with effects on gas transfer, bulk mixing and thus partly on growth and product formation: see Chain and Gualandi (1954), Steel and Maxon (1962). For a detailed review of product formation kinetics, see Pirt (1969).

In general, products as such do not lend themselves to control as a means of optimisation, but the existence of product effects may determine the optimal fermentation programme to be adopted. It may, for instance, be better to work at low overall concentrations in order to achieve high substrate conversions, even though fixed costs for the installation may support a trend to higher densities and maximum throughputs. The resolution of these conflicting effects on profitability may be made by separating the cost per unit of product into fixed and variable components for separate consideration. These may be recombined to give the lowest unit cost and thereby the optimal throughput rate.

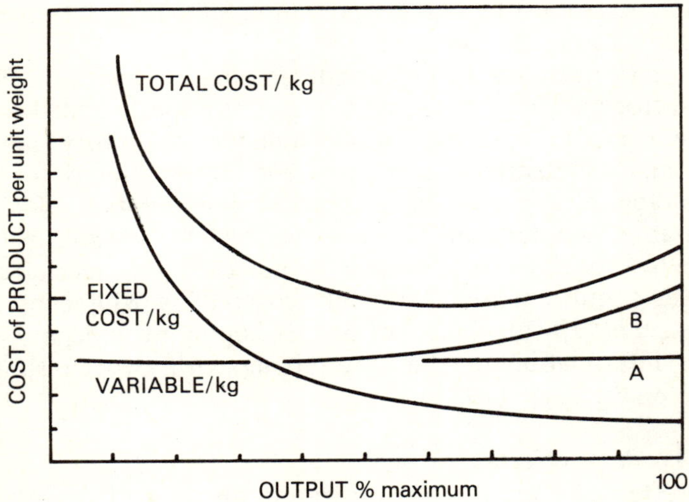

Fig. 5.2
Relationship of cost to throughput.

In Fig. 5.2 above, case A costs are independent of throughput rate, whilst in case B (typical of rather more installations, not just for continuous processes) the variable costs will rise (lower yields, more time lost on turn-rounds etc.) with increasing throughput, so that the resultant combined fixed and variable costs curve will have a definite minimum, *not* at the maximum throughput value.

When continuous measurement of product concentration is possible, a dynamic optimisation may be made for each batch run of a series, not as a means of controlling conditions in the run itself but as a way of deciding when the run shall be harvested. If the concentration of product is plotted against time, the first differential dY/dT is the production rate. If this in turn is plotted against time, the plot will denote change of productivity with time. This curve will have a distinct maximum for each run, as shown in Fig. 5.3.

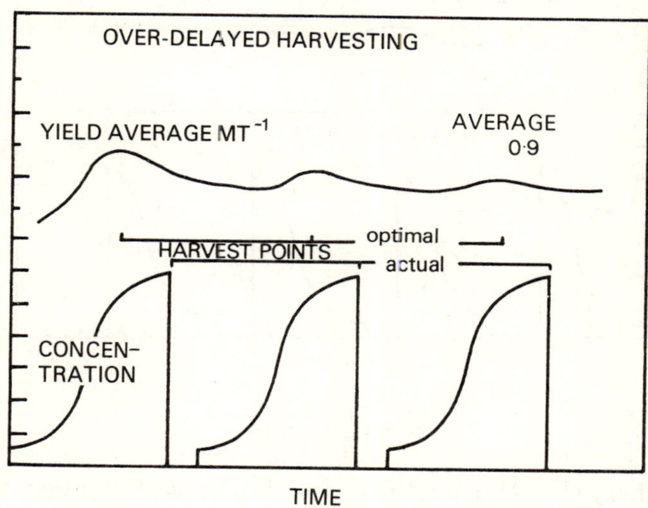

Fig. 5.3
Over-delayed harvesting.

When the value of productivity sinks down to such a level as to equal the *average productivity* of previous runs (including down time) the run should be stopped and the plant prepared for a new run. The updating of the value for average productivity is easier to obtain *via* on-line computing techniques, although these are by no means essential. This method is particularly valuable where runs vary in kinetics and no optimal harvesting point may be predicted beforehand. In Figs. 5.3 and 5.4 details are given of two systems, one harvested at almost standstill maximum yield, and the other with the harvesting optimised in the manner described. The data are hypothetical (from Vincent, 1972).

5.2.3 Physical and (non-nutrient) chemical variables

The dependence of growth and of product formation upon these external conditions may be the expression of effects upon individual enzymic reactions, which are often found to have optima in pH and temperature tending to coincide with growth conditions: see Gale (1940), Rabotnowa (1963), Jacob (1969), Patching and Rose (1969) and Munro (1969). Measurement of these variables will be dealt with later.

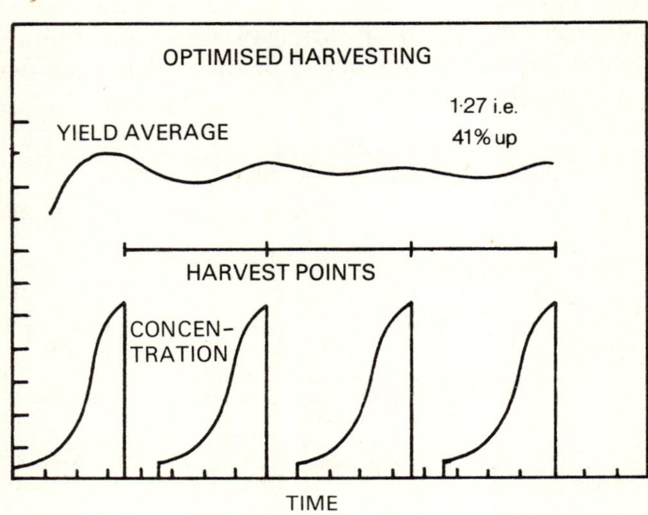

Fig. 5.4
Optimised harvesting.

Humphrey (1974) made the distinction between 'physical sensors' which measure the usual controllable variables such as temperature, motor speed, and foam level and those 'gateway sensors' which give information from which may be computed more fundamental facts about the fermentation, usually concerning the physical state of the culture. Humphrey's list is given in Table 5.1.

Table 5.1

'Gateway Measurements' in Fermentation

READING(S)	MAKES AVAILABLE BY CALCULATION
pH (properly titrant uptake†)	Acid/base formation
O_2 concentration difference in/out	
	O_2 uptake
Air flowrate	
	CO_2 evolution
CO_2 out	
O_2 uptake	
	Respiratory quotient

CO_2 evolution	
Power input	k_La
Airflow rate	
Coolant water use†	Heat evolution/metabolic energy
Product concentration†	Process kinetics and hence optimisation
Cell count†	Growth kinetics
Cell particle size distribution†	Phasing of cell division, synchrony

† additions by the reviewers

The importance of this concept of Humphrey cannot be overestimated. Only in fermentation systems which are not equipped with monitoring instruments up to or above a critical level is such a concept not of general utility. To reformulate the concept in our own way:

At and above a certain level of instrumentation of a process, the information available is sufficient to allow calculation of the values of fundamental process parameters in order to be able to describe the state of the process with considerably less uncertainty.

At the 'critical level' thus postulated, a qualitative improvement in knowledge of the process is achieved and we may expect to optimise the system in a more purposeful way, rather than empirically. Perhaps the criterion of having achieved this critical level of process information is whether or not mathematical models or algorithms may be written describing system behaviour which are representative of what happens during changes of system conditions. It will be evident that the critical level may be attained for different times: for instance, phenomena at the cell level may not affect bulk phenomena such as product formation over a long sampling period. The effects would be broadly similar in synchronised or asynchronous cultures.

Once, however, the critical level is achieved, we may develop to a much more accurate pitch, specific for each process, the dynamic 'optimisation' type of approach to batch culture illustrated in Fig. 5.4, and thereby lift the whole process to a high level of efficiency.

5.3 GENERAL CONSIDERATIONS IN SENSING, MEASUREMENT AND DISPLAY

The most important part of any measurement and control system is the sensor, since it is the primary source of information upon which subsequent electronic processing is dependent. In selecting sensors, properties such as accuracy, resolution, sensitivity and drift must be considered. These may be defined as follows:

Accuracy: the capability of the sensor to provide an analogue signal related to the true value of the stimulus (energy source) within defined probability limits. This is usually defined as a percentage of the range of the sensor.

Resolution: the smallest change in stimulus to the sensor which causes a significant change in output.

Sensitivity: the ratio of the change in sensor output to the corresponding change in the stimulus; e.g. a thermistor may have a sensivity of 20Ω/C.

Drift: the variation in the output of a sensor independent of change in the stimulus.

The better these specifications are, the better the measuring system will perform and the more expensive it will be. Generally, especially on the large scale where absolute values are of lesser importance, reproducibility is the more important criterion. This may be defined as the closeness of agreement between the output of a sensor measuring the same values of the stimulus under the same conditions at different times. This concept is particularly important when, for example, compensation has been applied to a signal in order to linearise it; the sensor or measuring instrument may not indicate the correct absolute value to within narrow accuracy limits, but provided that the reproducibility of the readings is good it is quite acceptable.

The basic function of a sensor or transducer is to convert one form of energy, e.g. light, heat, electrochemical or mechanical energy, into electrical energy. Ideally, the sensor would operate in such a manner that its output (in volts, amperes etc.) would be a linear function of the primary energy source; i.e. the system could be described by the equation for a straight line, $y = mx + c$. Whilst this is sometimes the case (e.g. pH and E_h are related to electrode potentials by the Nernst equation), there are many cases in which the output cannot be so accurately predicted and which can be arbitrary in nature; the sensor must then be matched to the signal conditioning and readout systems by calibration against input conditions of known absolute value. By doing this, it may be possible to relate the transducer output to the energy source in a manner which can be linearised electronically, even if only over a narrow working range. Examples of this type of behaviour are galvanic oxygen electrodes and strain gauges.

Electrochemical sensors are in general acted upon by changes in the chemical environment in which the sensor is situated; the sensor provides some electrical analogue of the causative variable. In addition to responding to changes in the main causative variable, the sensor will also respond to other stimulants in the system, of which the most pernicious is temperature. Temperature effects may in theory be eliminated if a fermenter is subject to rigorous temperature control, but this is not often the case. Some form of compensation of the sensor system is therefore usually a necessary attribute of the signal processing stage, and is considered later in this chapter.

A generalised type of measurement system for one variable only would have the following block layout (Fig. 5.5).

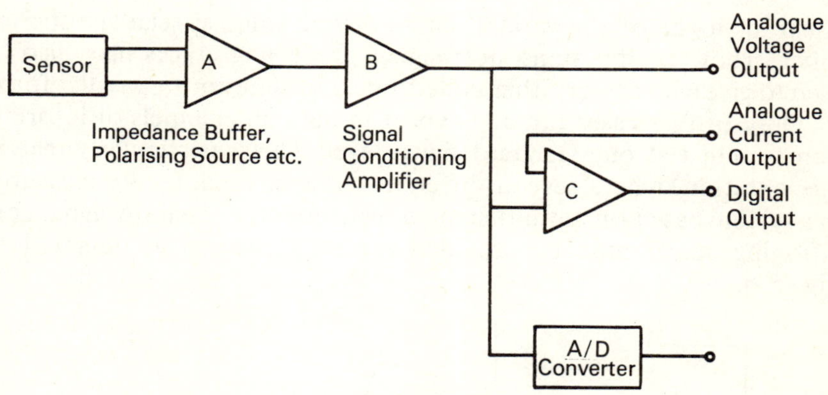

Fig. 5.5
Generalised measurement for one variable.

The sensor may generate its own current or potential (e.g. thermocouple, pH electrode, galvanic oxygen electrode), or may change its resistance as a function of the effect to be measured, and thus needs to be supplied with a constant source of current or voltage (e.g. resistance thermometer, resistance pressure/strain guage, dynamometer, polarographic electrode).

For simple measuring systems the current output circuit may not be provided. The voltage output would normally be used to drive indicating devices located either integral with the instrument or hearby. The current output is better for interference-free transmission over distances up to 300 metres; for distances longer than this, a digital conversion and data transmission system would be recommended.

Electrical signals may also be converted to air pressure analogues by means of an electropneumatic transducer (e.g. the Taylor-Sybron 701T, which accepts current or voltage inputs and delivers a linear air pressure signal normally in the range 0.1 – 1.0 bar, when supplied with air at 1.3 bar). The air pressure analogue may also be used to actuate remote displays or controller elements.

There is a general trend to standardise the outputs from transducers, so that as far as possible one type of signal conditioning instrument can be used for a wide range of different variables. This is particularly important where signals are to be fed to one central instrument such as a data logger, which in theory should be used to scan all incoming signals and then actuate printers at set intervals or only in alarm situations (e.g. the Leeds & Northrup Trendscan 1000 system). In spite of the trend to standardisation, however, signal levels from sensors or from the 'matching' stage (A in Fig. 5.5) may be mV d.c., 0 – 5 mA, 4 – 20 mA, 0 –1 V, 0 – 10 V, frequently a linear analogue of the variable or following some special law such as square root. Digital transducer/signal processing stages may also be used to give a binary or other coded signal in series or in parallel form.

Only in rare cases can one expect to have all channels of information supplied at one standard signal level. This is particularly true in fermentation, where several different kinds of variables are measured In a system based on a multichannel instrument, a separate signal conditioning stage must be provided for each channel as depicted in Fig. 5.6.

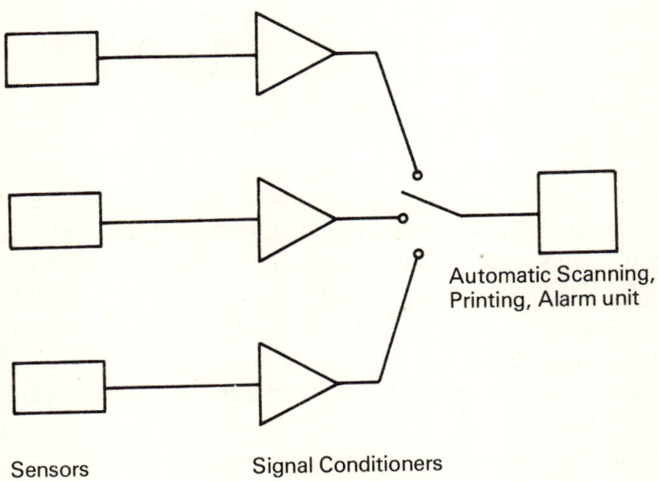

Fig. 5.6
Generalised scheme showing signal conditioning before presentation to scanner.

5.3.1 Display of signals

There are two fundamental methods for indicating current or voltage derived from a transducer to give an instantaneous display: analogue and digital. Analogue display is still preferred by some plant operators since a signal change or trend from steady state can be seen at a glance from the position of a needle on a dial, whereas such facile interpretation is impossible with a digital indicator. As a general rule, analogue meters are to be preferred when highly accurate data are not necessary; one must always bear in mind that a digital meter reading to two places of decimals (e.g. for readout of pH) is of little advantage if the rest of the measuring system is accurate to only 5% at best. In the plant, analogue indication at source, with digital data logging at a central station, approaches the ideal.

Analogue meters still have an advantage over digital versions in cost terms, but this advantage is likely to disappear as the cost of the displays and drivers for digital circuitry falls with increasing production, while the cost of manufacturing high precision mechanical components of the type found in analogue meters rises. Digital meters have a further advantage in that they are less sensitive to vibration and shock, and can be mounted in any position.

5.3.1.1 ANALOGUE METERS

The most common form of analogue meter in use in the moving coil galvanometer, and discussion will be confined to this type. They are available in a wide range of shapes and sizes, scale lengths, specifications and modes (although primarily current-reading instruments, they are available with integral multipliers for the reading of voltage).

It should be noted that manufacturers' specifications of accuracy, resolution and drift are invariably referred to the range or full-scale deflection (FSD) of these (or indeed any other) measuring instruments, usually as a percentage. Thus an instrument with a range of 0 – 100 μA and a quoted accuracy of 0.01% will read a signal of 100 μA to 0.01 μA. For an input of 10 μA the accuracy is still 0.01 μA, which is in fact 0.1% of the input. Similarly, a signal of 1 μA will only be read to 1% of the input. Similar reasoning can be extended to specifications of resolution and drift.

Current-reading instruments may easily be converted to read voltage by connection in series with a multiplier or potential divider network, the negative terminal of the meter being connected to ground or the negative side of the input signal.

5.3.1.2 DIGITAL METERS

The main advantage of digital meters compared to their analogue counterparts, other than the obvious benefit in some cases of presenting an analogue signal in digital form, is their high input impedance; this is usually of the order of 10 MΩ, compared to values of the order of 1500 Ω for analogue meters. It is therefore possible in some cases to present the output signal of a transducer directly to the meter without interfacing buffering circuitry. This does, of course, depend upon the matching between the mode and input range of the meter and that of the output of the transducer.

Digital meters consist essentially of power supply, analogue-to-digital conversion circuitry, and the display and its associated driving circuitry. Standard scales are 0 – 99 (2 digit), 0 – 199 (2½ digit), 0 – 999 (3 digit), 0 – 1999 (3½ digit), 0 – 9999 (4 digit), and 0 – 19999 (4½ digit). Most provide for variable ranges by means of internal attenuating resistors, so that for example one instrument can read 9.99, 99.9, or 999 mV FSD. Depending upon the interfacing circuitry, these digits can be scaled to read amperes, volts, MHz, or even in engineering units such as pH. Displays are normally light-emitting diode (LED) or liquid crystal (LC); the LC is becoming increasingly popular as prices fall, and because of its extremely low current consumption which renders it ideal for use in battery-powered equipment.

As is the case with analogue meters, specifications must be interpreted with care. Accuracy is normally specified in terms of % FSD or ±1 digit, whichever is the larger. Thus, an input of 998 mV could be displayed as 997 to 999 mV, an accuracy of approximately 0.2%.

Other facilities which may be available in digital meters are bipolarity (useful for indication of E_h) and decimal point switching. The latter facility is especially useful for display in engineeting units, especially if the meter is to be used on a time-shared basis for the display of more than one variable.

5.4 METHODS OF MEASUREMENT

All the systems described in this section use some sort of operational amplifier for signal conditioning and amplification, and produce a signal suitable for driving a panel meter. It is therefore useful at this stage to summarise the main properties of such amplifiers.

5.4.1 Integrated circuit amplifiers

There is an almost bewildering range of integrated circuits available on the market which can be used as amplifiers in linear circuits. It is not the intention of the reviewers to go into details of circuit design, a subject which has been competently covered by others (e.g. Hnatek,

1975; Driscoll and Coughlin, 1975); a brief summary of the function of operational amplifiers may be useful, however, and is relevant to subsequent discussion.

Two fundamental configurations of an operational amplifier are shown in Figs. 5.7a and 5.7b.

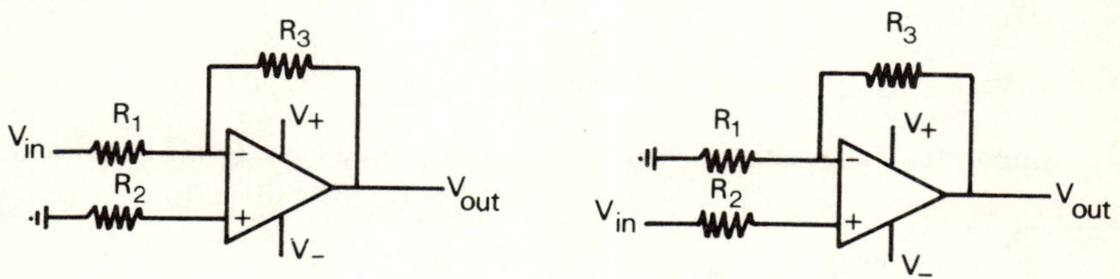

Fig. 5.7

(a) Inverting amplifier circuit. (b) Non-inverting amplifier circuit.

In Fig. 5.7a the amplifier operates in the inverting mode, that is the output signal is of the same magnitude (ideally) and of opposite polarity to the input signal. The gain of the amplifier is defined as:

$$G = \frac{-V_{out}}{V_{in}} = \frac{R_3}{R_1} .$$

In Fig. 5.7b the amplifier operates in the non-inverting mode, that is the voltage out is of the same polarity as the input voltage. In this case the gain is defined as:

$$G = \frac{V_{out}}{V_{in}} = \frac{R_1 + R_3}{R_1} .$$

Most amplifiers are used in this negative feedback mode, which has the effect of improving gain stability, improving linearity, reducing output impedance and (in some cases) increasing input impedance. If the gain is sufficiently large, the characteristics of the closed-loop amplifier become a function only of the feedback components.

In theory, the ideal unity gain amplifier will have an output voltage numerically equal to the input voltage. In practice this is not so, owing to imperfections in the device itself. The most important of these are:

Input bias current: the difference between the currents flowing in the input terminals at zero output voltage.

Input offset current: the difference between the currents flowing in the input terminals at zero output voltage.

Input offset voltage: the voltage which must be applied through two equal resistances to drive the output voltage to zero. Most operational amplifiers provide terminals for nulling the offset voltage.

Typical values of these parameters are given in Table 2 for four common integrated circuits.

Table 5.2

Important characteristics of some commonly used operational amplifiers

PARAMETER	DEVICE			
	LM741[a]	LM308[a]	LM725[a]	NE536T[b]
Input impedance (Ω)	10^6	40.10^6	$1.5.10^6$	10^{12}
Output impedance (Ω)				100
Bias current (nA)	200–500	1.5–7.0	42–100	5 pA
Input offset current† (nA)	30–200	0.2–1.0	2–20	
Input offset voltage† (mV)	1.0	0.3	0.5	
CMRR (dB)	90	110	100	

† source resistance 10 kΩ
(a) source: National Semiconductors Linear IC Data Book
(b) source: Signetics Technical Handbook

These devices will all source a current of 25 mA, which is sufficient to drive most indicating instruments.

5.4.1.1 COMMON MODE REJECTION RATIO (CMRR)

Common mode interference is a voltage, usually at mains frequency, induced in the terminals of a signal circuit with respect to a reference point (usually earth). It is common to all terminals of the circuit, and is exacerbated by imbalances in the circuit (e.g. in the leads to the amplifier itself) and by imbalances on the input pins of the amplifier (e.g. if $R_1 = R_2$ in Figs. 7a and 7b, and if the capactances from the input pins to ground are not matched). The capacity of the amplifier to operate in the presence of common mode interference is called the **Common Mode Rejection Ratio**, and is defined as the ratio of the input voltage range to the change in input offset voltage over this range. It is usually specified in decibels.

5.4.1.2 INSTRUMENTATION AMPLIFIERS

Discussion so far has been based on the assumption that the sensor signal is transmitted over a relatively short distance (2 metres) to the amplifier. The open-ended systems used in the illustrations, i.e. in which one input terminal is referred to ground, are adequate for processing such signals without undue electrical interference. However if the sensor signal is to be transmitted across the plant for a considerable distance the noise voltages induced on the cables would be excessive, and would swamp the signal. The use of a differential amplifier, which accepts both the positive and negative sensor signals, would be required in this case. Instrumentation amplifiers, with differential inputs, are precision amplifiers designed for this purpose. Even so, considerable difficulties may still remain, and it is usually necessary to guard the signal source or the amplifier. A guarded amplifier system, in which the amplifier input stage is isolated from the rest of the signal processing circuitry, is the usual choice; the guard is connected to the remote transducer earth. Such a system considerably enhances the capacity of the amplifier for common mode rejection.

5.4.1.3 TEMPERATURE COMPENSATION

Temperature compensation of sensor outputs is most easily effected at the amplification stage by either manipulating the gain or altering the biasing of the input terminals, and may be done either manually or automatically. Manual compensation systems can take the form of a variable resistor inserted into the gain loop, the value and mode of the resistor (linear or logarithmic) being selected so as to modify the gain, based on a knowledge of the temperature coefficient of the input signal. A typical example of this is the temperature compensation potentiometer on some pH meters, which is calibrated in °C; such a system would then be compensated for the selected temperature only.

Automatic compensation involves the use of circuit elements which change in a known manner with temperature, such as thermistors (see Section 5.4.2.1). A thermistor network, or preferably a linearised resistor/thermistor network, will reduce the gain in response to temperature rises and *vice versa.* A simple arrangement of this type is shown in Fig. 5.8.

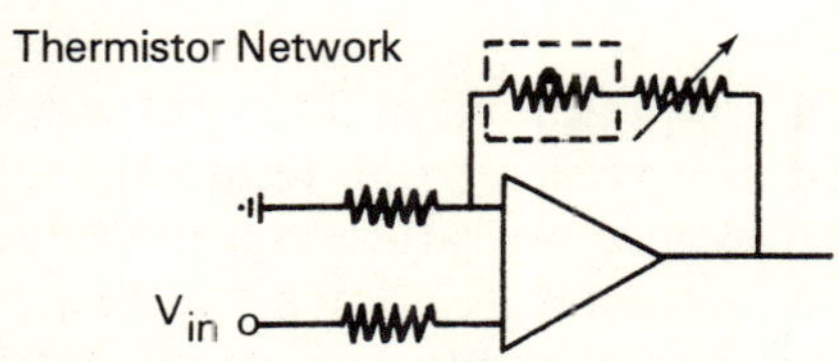

Fig. 5.8
Compensation of amplifier for temperature coefficient of signal, using thermistor network.

The thermistor network would, of course, be mounted remotely from the amplifier, and may well be integral with the sensor.

Other methods of temperature compensation are possible, such as the amplified and conditioned signal from a Pt100 sensor; this could operate on the principle of altering the biasing of the amplifier. For an example, see Fig. 5.13 in Section 5.4.3.5. Temperature compensation may also be effected by comparing the output of the sensor amplifier with a linear temperature analogue in a differential ampli-

5.4.2 Measurement of temperature

5.4.2.1 THERMISTORS

Thermistors are semiconductors which exhibit a large change in resistance for a relatively small change in temperature. They may be obtained in the shape of beads, rods, discs and probes (i.e. encapsulated in a stainless steel or similar sheath). Bead types are generally used for temperature measurement, whether sealed in a glass envelope or mounted on an alloy disc. A general guide to temperature ranges of application is shown in Table 5.3.

Table 5.3

Resistance and temperature ranges of thermistors

TEMPERATURE RANGE (°C)	TYPE OF THERMISTOR	RESISTANCE RANGE AT 25°C
–70 – +45	Low resistance	100Ω – 2 kΩ
45 – 150	Medium resistance	2 kΩ – 75 kΩ
150 – 320	High resistance	75 kΩ – 500 kΩ

For temperature measurements in fermentation the low-resistance type is therefore used.

Thermistors with few exceptions, have negative temperature coefficients. The resistance-temperature relationship is non-linear, and takes the form:

$$R \propto e^{B/T}$$

Thermistor data are presented in different ways, for example: the B value and a resistance value at one temperature; as plots of log R versus temperature; as a temperature coefficient; or as direct charts of resistance versus temperature. Since the response of thermistors to temperature is non-linear, the temperature coefficient of resistance (sensitivity or slope) varies with temperature; in general, the higher the temperature coefficient of resistance the higher is the characteristic temperature (B value).

Linearised thermistor networks (thermistor/resistor composites) are combinations of high-precision resistors and thermistors, and are used to overcome the intrinsic non-linearity of thermistor response to temperature. They are connected in networks which produce a varying voltage or resistance linear with temperature. Typical slopes of linearised networks are: voltage mode up to 20 mV°C; resistance mode 17 – 130 Ω/°C.

Self heating is caused by current passing through the thermistor generating heat as I^2R_t. As long as self heating is negligible the thermistor assumes the temperature of the ambient. The rate of temperature rise is determined by the thermal time constant, which is the time required for the thermistor to change temperature by 63% of the ambient temperature rise. Self heating must therefore be avoided in temperature measurement, although it is used in flow measurement and other techniques. The thermal time constants of thermistors used for measuring temperature must be of the order of one second or less.

In practice, thermistors are rarely used for direct temperature measurement, but are extensively used in temperature compensation circuitry (see Section 5.4.1.3).

5.4.2.2 RESISTANCE THERMOMETERS

A resistance thermometer is a temperature-sensitive resistor which produces a voltage change when a current is passed through it. A platinum wire of 100 Ω resistance (Pt100) is conventionally used, although other resistance values are available. The resistance-temperature relationship for a platinum resistance thermometer is non-linear, and is represented by the equation:

$$R_T = R_0\ (1 + AT + BT^2 + CT^3 + \ldots\ldots)$$

where R_T is the resistance at temperature T, R_0 is the resistance at some reference temperature, usually 0°C, and A, B, and C are constants.

Over a limited temperature range, such as is used in fermentation, a linear approximation of this equation may be assumed.

$$R_T = R_0\ (1 + AT)\quad .$$

A simple bridge circuit based on this assumption of linearity is shown in Fig. 5.9.

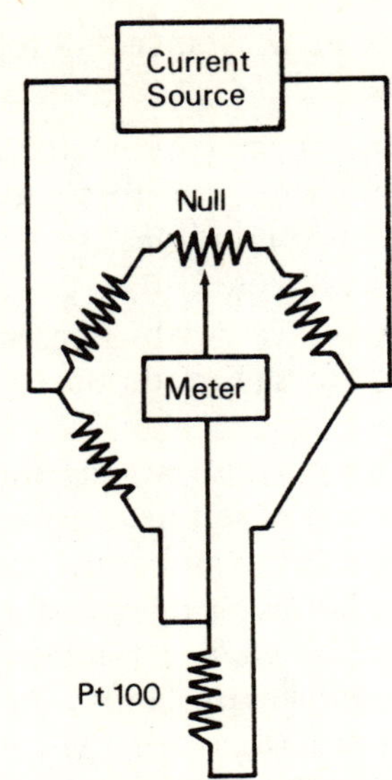

Fig. 5.9 Typical circuit for thermocouple.

Commercial Pt100 amplifiers use more complex circuitry to linearise the output of the sensor over a wider range, but this will not be treated here.

5.4.2.3 SEMICONDUCTOR JUNCTION SENSORS: DIODES

Like all semiconductors, diodes are temperature-sensitive; this property can be used to measure between about –20°C and +150°C. As is the case with the Pt100 system, a stabilised current source is required. A simple circuit for such a sensor is given in Fig. 5.10.

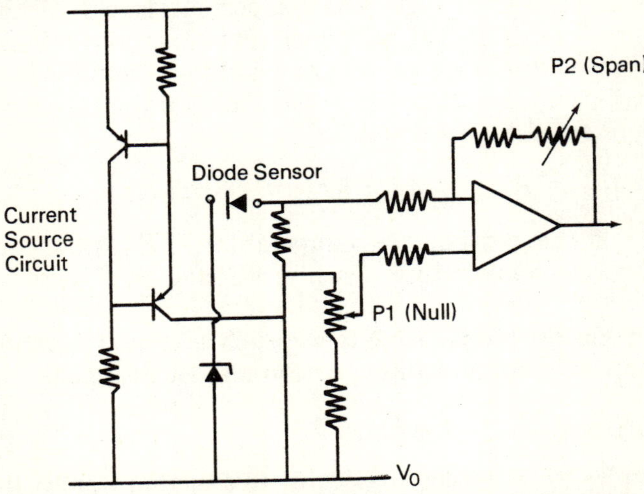

Fig. 5.10 Temperature measurement circuit using diode sensor.

The amplifier output will be essentially stable with temperature; potentiometer P1 is adjusted so that the output of the amplifier is 0 mV at 0°C, and then the scaling potentiometer P2 is adjusted so that the output is for example 1000 mV at 100°C. The signal can then be fed to a measuring instrument.

5.4.3 Measurement of pH

The measurement of pH is a well-established technique in both laboratory and industrial plant. Consequently, equipment has reached a stage of high refinement, which has been exemplified in recent years by the adoption of the combined glass/reference electrode and by the use of integrated circuits with very high impedance input stages (FET operational amplifiers) in the measuring circuitry. It is now possible to measure pH with the same accuracy and stability as was obtained with valve voltmeters and electrometers previously essential for this work, at a considerably lower cost.

5.4.3.1 FUNDAMENTALS OF MEASUREMENT

In measuring pH we are in reality measuring the hydrogen ion activity. These two properties are related by the expression:

$$\mathrm{pH} = -\log a_{\mathrm{H}^+}$$

The electrochemical measurement of a_{H^+} is governed by the Nernst equation, which in this case takes the form:

$$E = E^{\circ} + kT.\mathrm{pH} \qquad (5.1)$$

where E is the measured electrode potential; E° is the standard (reference) electrode potential; kT is the slope factor = $2.303RT/\mathrm{F}$, where R is the Universal Gas Constant, F is Faraday's Constant, and T is the temperature (°K).

5.4.3.2 TEMPERATURE EFFECTS ON pH MEASUREMENT

Temperature effects are the main source of error in pH measurement, and are caused by the temperature-dependence of the slope factor and the reference potential, the latter being manifested as zero shift error.

Slope factor: a table of slope factors in terms of mV/pH unit as a function of temperature is given as Table 5.4.

Table 5.4

Variation of slope factor with temperature

T (°C)	SLOPE FACTOR (mV/pH)
0	54.196
5	55.188
10	56.180
15	57.172
20	58.164
25	59.156
30	60.148
35	61.140
40	62.132
45	63.124
50	64.116
55	65.108
60	66.100
65	67.092
70	68.084

(from Munro, 1970)

Zero shift error This applies only if the reference electrode is situated remotely from the glass (measuring) electrode (e.g. outside the fermenter). The standard electrode potential E° of the reference cell is temperature-dependent; and if the glass and reference electrodes are at different temperature an error is introduced. This source of error can be eliminated by the use of combined glass and reference electrodes.

To analyse this phenomenon it is necessary to introduce the concept of isopotential, which is the pH value at which the measured electrode of an electrode pair is approximately constant over a limited temperature range. This is shown in Fig. 5.11.

The dependence of the reference electrode potential E° on temperature can be expressed as:

$$E^\circ = E^\circ_o + AT$$

where E°_o is the reference electrode potential independent of temperature, and A is a constant. Substituting this expression into Equation (5.1) (Section 5.4.3.1):

so $$E = E_o^o + AT + kT.\mathrm{pH}$$

so $$\frac{dE}{dT} = A + k.\mathrm{pH} \quad .$$

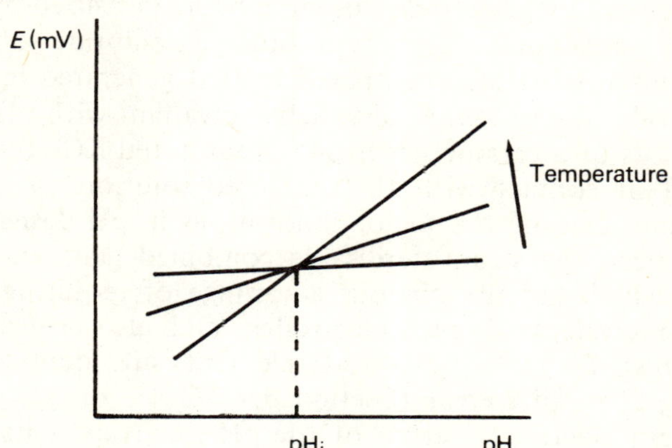

Fig. 5.11
Relationship of electrode potential to pH as a function of temperature, showing isopotential pH ($\mathrm{pH_i}$).

At the isopotential point $\mathrm{pH_i}$, $dE/dT = 0$, so that:

$$A = -k.\mathrm{pH_i} \quad .$$

Thus $$E = E_o^o + kT(\mathrm{pH} - \mathrm{pH_i}) \quad . \tag{5.2}$$

The term $kT.\mathrm{pH_i}$ is the zero shift error.

To obtain true compensation for temperature the concept of isopotential must be extended to the glass electrode also. From Equation (5.2):

$$\frac{dE}{dT} = kT \frac{d\mathrm{pH}}{dT} + k.\mathrm{pH} - k.\mathrm{pH_i} \quad .$$

At the isopotential point:

$$(\mathrm{pH_i})_c = \mathrm{pH_i} - T_{ref}. \frac{d\mathrm{pH}}{dT}$$

where $(\mathrm{pH_i})_c$ is the pH of the combined measuring and reference systems, T_{ref} is the reference temperature, and $d\mathrm{pH}/dT$ is the temperature coefficient of the sample.

5.4.3.3 pH ELECTRODES

The pH electrode has been in common use in laboratories for many years, and needs little further description. The electrode basically consists of a bulb of pH-sensitive glass fused to the end of a highly-insulating glass tube, electrode potential being measured by means of a Ag/AgCl or calomel electrode in buffered chloride solution. The electrode potential is then compared to that generated by the reference electrode, the potential of which is invariant with pH. This normally consists of a calomel electrode in saturated KCl, the cell being maintained in contact with the measured solution by means of a ceramic plug. One of the major innovations in pH design in recent years has been the steam-sterilisable combined glass and reference electrode, which has the obvious advantage of requiring less space than the equivalent discrete electrodes, and also ensures that the temperatures of the two electrode elements are identical, thereby eliminating zero shift error (Section 5.4.3.2).

The main electrical feature of the pH electrode is its extremely high output impedance, which is of the order of $10^8\Omega$. This necessitates the use of special cable and plugs to connect it to the measuring circuit, especially if the whole electrode is to be sterilised in an autoclave. The main disadvantages of pH electrodes are an irreversible increase in output impedance caused by steam sterilisation, and fragility. Some progress is being made with the latter by using ceramic materials (e.g. the Pfaudler electrode), but the major obstacle is to provide enough resistance in the stem to insulate both sides of the glass bulb.

The output of a pH electrode pair is zero at one particular pH value, normally pH7 (although sometimes pH2 or pH4). That is important in the design of pH amplifiers (Section 5.4.3.5) in that at this E^o value the polarity of the output signal will reverse. An inverting amplifier is therefore necessary in the measuring circuitry in order to generate an output capable of being displayed in pH units.

5.4.3.4 ELECTRODES FOR MEASURING OTHER PARAMETERS BY MEANS OF pH CHANGE

5.4.3.4.1 pCO_2 electrodes: The measurement of partial pressure of dissolved CO_2 is possible with a pH electrode, since the dissolution of CO_2 in an aqueous medium is accompanied by a pH shift. The pH change is extremely small, so that an amplifier with a scale expansion facility (+0.5 pH) and accurate temperature compensation is required.

pCO_2 electrodes basically consist of a combined pH electrode with a bicarbonate buffer surrounding the bulb and ceramic plug, the solution being retained by means of a thin (5μm) PTFE membrane secured by means of an O-ring (e.g. the Probion electrode).

The response times of pCO_2 electrodes are very slow, especially to a negative step change in partial pressure; this leads to considerable time lags in the measurement system, so that it is impracticable at the time of writing to use such electrodes in automatic control systems.

5.4.3.4.2 Enzyme electrodes: Any enzymatic reaction which produces a change in pH can be used as the basis of an enzyme electrode. One such experimental electrode for measuring penicillin concentration is described by Nilsson and Mosbach (1978). In this system, a small volume (50 μl) of lactamase was held against the bulb and ceramic plug of a pH electrode by means of a dialysis membrane. As is the case with pCO_2 electrodes, change in pH accompanying such reactions are extremely small, and amplifiers with expanded scale facilities and temperature compensation are essential.

Enzyme electrodes are still in an early stage of development and have been included in this section solely to show the range of measurment possibilities which will be available in the future.

5.4.3.5 AMPLIFIERS FOR pH ELECTRODES

Since the source impedance of a pH electrode is of the order of $10^8 \Omega$, the first stage in any pH measurement system must be an opertional amplifier with an even higher input impedance. The development of FET and CMOS technology has made available operational amplifiers with input impedances in excess of $10^{12} \Omega$.

A commonly used device is the Signetics NE536T (see brief specification in Table 2). This is used as a low gain impedance buffer between the electrode and subsequent signal processing circuitry. Since the output impedance of the buffer is low, subsequent amplification can be performed by devices of relatively low input impedance such as the LM741 or LM748. A very simple circuit with no temperature compensation is shown in Fig. 5.12.

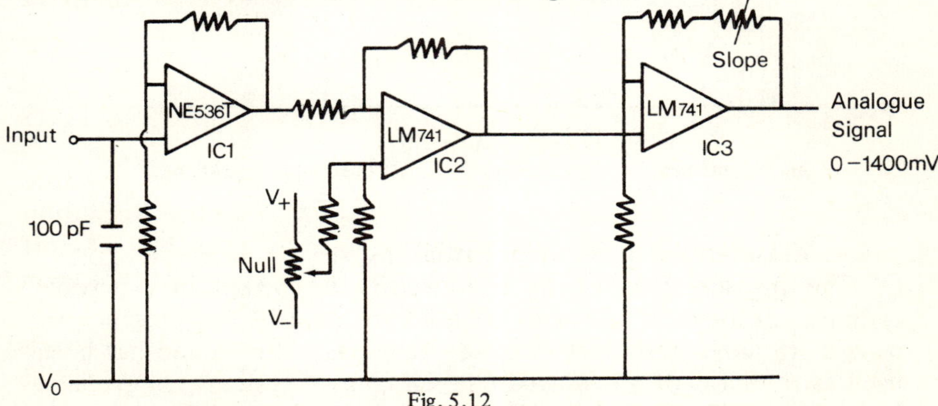

Fig. 5.12
pH measurement circuit using NE536T FET operational amplifier.

Points of interest in this circuit include: (a) the 100 pF capacitor in parallel with the input to damp out induced voltage on the input pin; (b) the low gain input stage – $G = 2$; (c) the inverting amplifier (IC2); (d) the null (buffer) potentiometer, which injects a signal to IC2, which then inverts the difference; (e) the slope potentiometer is, in effect, a form of manual temperature compensation in that if the output is scaled at, say, 30°C, this will be equivalent to the true pH only at this temperature.

This circuit can be used either with an analogue meter with a left-hand zero, or with a digital meter. If a centre-zero meter is used in combination with a pH electrode with an E^0 at pH7, the inverting amplifier would not be necessary.

This is a relatively simple system; Fig. 5.13 shows an alternative, using a CMOS device (RCA CA3130). This has facilities for both manual and automatic temperature compensation, the latter being effected by means of a signal from a Pt100 source, which alters the biasing of the buffer.

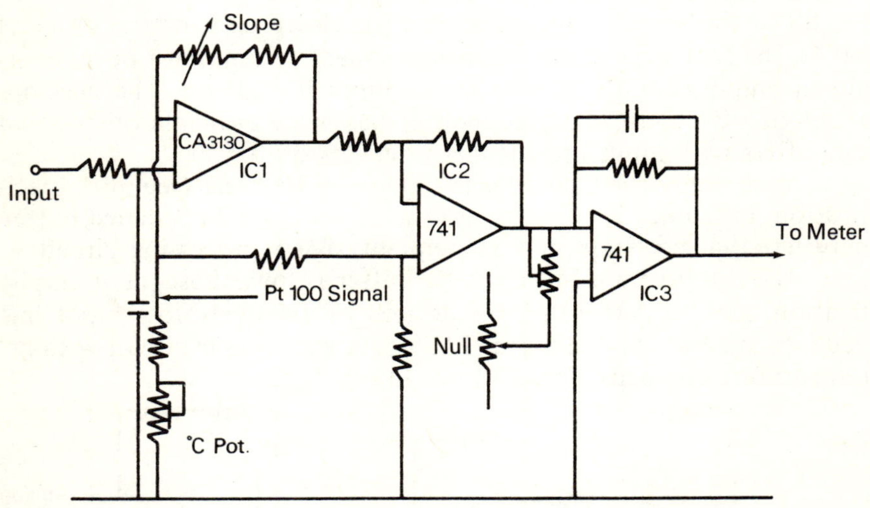

Fig. 5.13
pH measurement circuit using CA3130 CMOS operational amplifier.

5.4.4 Measurement of redox potential (E_h)

The measurement of E_h is especially important in anaerobic systems, although it can also be used to indicate the presence of oxygen (Squires and Hosler, 1958; Tengerdy, 1961), and has been used as the basis of a dissolved oxygen control system (Lengyel and Nyiri 1965). The technique of measuring E_h is essentially similar to that for pH, since both are governed by the Nernst equation; E_h is therefore also temperature-sensitive. Since the measured potential can be either positive or negative, the conditioned signal should either be fed to a bipolar digital meter or a centre-zero analogue meter; alternatively, a null facility could be provided with sufficient back-off potential to render the signal positive over the whole range. A typical circuit for E_h measurement is shown in Fig. 5.14.

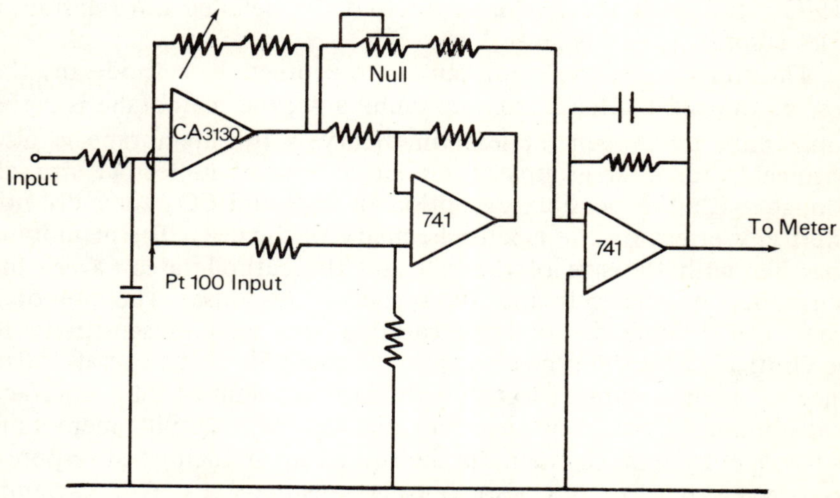

Fig. 5.14
E_h measurement circuit (based on that shown for pH measurement in Fig. 5.13).

It should be noted that E_h varies with pH; this is discussed in detail in terms of rH by Jacob (1970).

5.4.4.1 REDOX ELECTRODES

These consist of a noble metal cathode, normally smooth platinum, fused to the end of a glass tube. The size and shape of the cathode do not affect the magnitude of the output signal. The cell circuit is completed by a reference electrode of the same type as is used in pH measurement.

5.4.5 Measurement of dissolved oxygen

Many techniques are available for the measurement of dissolved oxygen, including chemical analysis (the Winkler method). We are concerned here, however, only with those of direct relevance to fermentation and similar studies; evidently, the measurement of dissolved oxygen by chemical methods for a system containing many fermenters from which periodic and regular (if not continuous) pO_2 readings are required would necessitate much labour on the part of laboratory staff, even with automatic equipment, and such measurements would be of no use for direct feedback control purposes. Hence, only those measurement techniques involving the use of electrochemical sensors will be considered here. Of these, the most important are the membrane oxygen electrodes.

The theory of the membrane oxygen electrode has been reviewed by Brown (1970), Vincent (1974a) and by Vincent and Priestley (1975), to which the reader is referred for detailed information; a brief summary only is given here.

The function of the membrane is to protect the cathode and the rest of the cell contents from contaminants; the membrane is permeable only to oxygen (at least in theory – the membrane is also permeable to other groups of similar or smaller molecular size, although with the possible exception of H_2S and CO_2 they do not exert any effect on the electrochemistry of the cell). The membrane together with the size of the cell and the cathode area govern the sensitivity of the cell and its speed of response. The use of a small cell volume and a large cathode area enables sensitivity to be controlled by the type of membrane used. Since the overall resistance to oxygen transfer to the cathode is the sum of the membrane and boundary layer resistances, a thin, high permeability membrane will concentrate the resistance in the liquid film, inducing fast response to oxygen tension but also causing sensitivity to hydrodynamic effects. A thicker, lower permeability membrane will concentrate resistance within the membrane, eliminating this stirring effect; the oxygen tension in the bulk of the liquid will be proportional to the current output of the electrode, but the response time will be dependent on the thickness of the membrane.

For continuous measurement and control purposes, the permeability of the membrane material must be selected so as to optimise response time with minimal hydrodynamic effects. In order to minimise the oxygen consumption of the electrode its output current must be low; amplification of the signal is therefore needed prior to visual indication.

5.4.5.1 TYPES OF MEMBRANE OXYGEN ELECTRODES

There are basically two types: polarographic (applied potential) and galvanic (self-polarising). Polarographic electrodes operate on the principle of electrochemical reduction of uncombined oxygen at a cathode (normally Ag) negatively polarised with respect to a reference anode (normally Pt or Au); the magnitude of the current difference between the zero potential and the discharge potential of oxygen (approximately 0.75 V) is proportional to the rate of cation transport to the cathode. The current output therefore gives a direct indication of the dissolved oxygen tension for the fixed system geometry used.

Galvanic electrodes rely on the potential difference generated between two dissimilar metals (e.g. Ag/Pb, Al/Ag) to reduce the oxygen; this results in the gradual dissolution of the less noble metal. Galvanic oxygen electrodes have now been almost universally adopted because of their lower cost, although their life is considerably less than that of polarographic electrodes. The cost of the signal processing circuitry is also less, since a polarising source does not have to be provided.

5.4.5.2 TEMPERATURE EFFECTS WITH MEMBRANE ELECTRODES

The effect of temperature on the cell current is given approximately by the equation:

$$I_T = A\mathrm{e}^{-J/T}$$

(Briggs 1971), where A is a constant and J is the characteristic temperature of the membrane material in °K. The permeability of the membrane is also temperature-dependent. As shown in Section 5.4.5.1, if the cell has a relatively large cathode area and a small cell volume, most of the resistance to oxygen permeability will be concentrated in the membrane; the thinner the membrane, therefore, the more pronounced will be the effects of its temperature coefficient.

A third temperature-dependent factor is introduced if measurements are required in terms of dissolved oxygen *concentration* (p.p.m. or mg/l) as opposed to partial pressure (% saturation). This is caused by the negative temperature coefficient of oxygen solubility, which is of the order of 2%/°C.

These temperature effects are normally suppressed by the use of a thermistor network. Although at first sight it would appear that the cell could be compensated by selecting a thermistor with a characteristic temperature equal to that of the membrane material, this is impossible in practice since the thermistor resistance would be too large. Compensation should therefore be executed in the amplifier, not in the electrode (see Fig. 5.8, Section 5.4.1.3 and Fig. 5.15, Section 5.4.5.3).

It may also be necessary to compensate for the salinity of the sample (by salinity is meant the presence of nutrients and extracellular products which will reduce the solubility of oxygen; they need not be saline in nature). This is much more difficult, since the effect of salts etc. on oxygen solubility will of course depend upon the composition of the sample. The relationship is generally linear, but gravimetric tests should be carried out if linearity is in doubt. What we have to provide here, therefore, is additional adjustment for the negative oxygen solubility coefficient; this cannot be done automatically, and therefore a manual adjustment would have to be provided in the measuring circuitry for say 0 – 35% salinity by means of attenuation of the amplifier output. The total salinity would then be determined gravimetrically, and the salinity adjustment on the amplifier would be used in conjunction with a calibration chart.

5.4.5.3 AMPLIFIERS FOR USE WITH DISSOLVED OXYGEN ELECTRODES

This discussion will be confined to amplifiers for use with galvanic electrodes, since these are in most common usage. Amplifiers for use with polarographic electrodes are essentially similar, but differ in the need for providing a stabilised voltage supply.

Amplifiers for use with galvanic electrodes are relatively simple, since the signal has a low output impedance (c.f. pH electrodes, Section 5.4.3.3). The current output of the electrode (normally of the order of 0–15 μA for 0–100% saturation, depending upon the electrode configuration and temperature) is normally dropped across a resistor, either in the electrode itself (to obviate the need for a shorting plug during sterilisation) or across the input of the amplifier. A 1 kΩ resistor will therefore result in an input signal to the amplifier in the range 0–15 mV. The extent to which this signal is to be amplified depends upon the method to be used for indication, and any attenuation thought to be necessary for temperature compensation. The circuit shown in Fig. 5.15 would be suitable.

The amplifier has to perform two main functions: it has to provide for suppression of the base current of the electrode, i.e. the current generated by the electrode in a zero oxygen environment; and it has to have a variable scaling facility in order to set the meter or other indicating instrument to full scale when the electrode is in a saturated oxygen environment. These functions are provided by the potentiometers RV_1 and RV_2 respectively.

The null potentiometer RV_1 therefore has to be capable of injecting a stabilised voltage of approximately –3 mV onto pin 3 of the amplifier; the span or 100% potentiometer RV_2 has to be capable of providing enough resistance, in combination with the other resistor elements in the feedback loop, to set the maximum gain at about 1000; the maximum output signal would then be 1500 mV, suitable for indication on an analogue or digital panel meter.

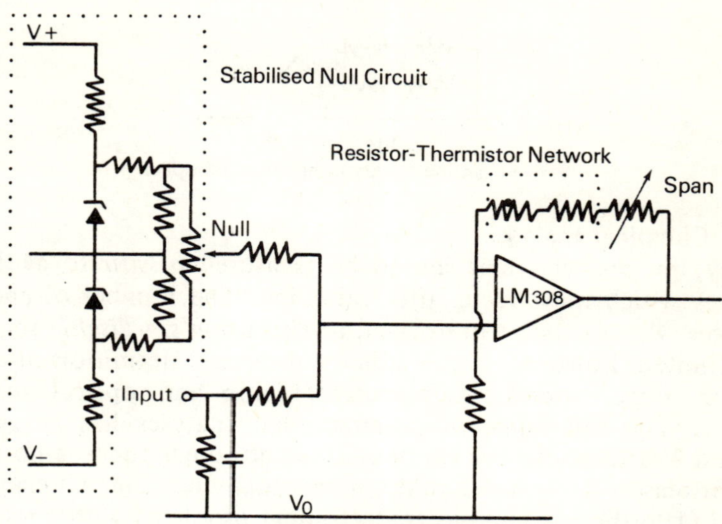

Fig. 5.15
Simple pO_2 measurement circuit for galvanic oxygen electrode.

It is often useful to provide an expanded scale facility on meters for indication of dissolved oxygen or pH; e.g. scales of 0 – 10% and 0 – 30% FSD in the case of dissolved oxygen. This can be accomplished by adding another amplifier in series, with switchable feedback loops of different gain. The first amplifier in this two-stage system would then be used as a low-gain buffer, the main amplification being executed in the second device; the different feedback loops would have their resistors tuned to give FSD for inputs corresponding to 10% and 30% saturation depending upon the position of the switch. An alternative method of scale expansion is to use a switchable series multiplier network following a buffer amplifier. These two systems are shown in Figs. 5.16 and 5.17 respectively.

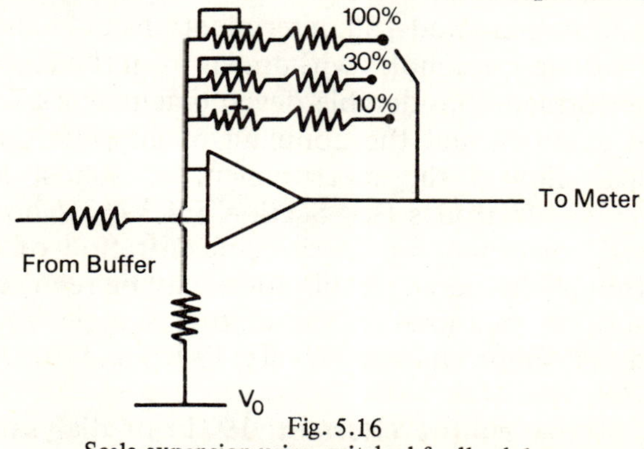

Fig. 5.16
Scale expansion using switched feedback loops.

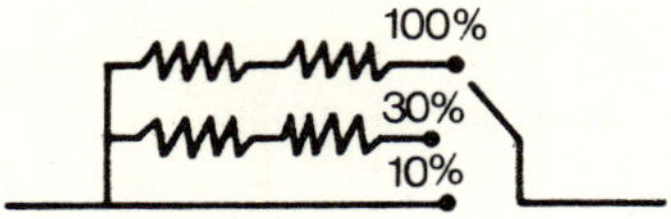

100%
30%
10%

Fig. 5.17
Scale expansion facility using series multiplier.

5.4.6 Chemical analyses

Ideally, process variables should be measured *in situ* to avoid any changes which may occur after sampling. The number of chemical analyses which may be performed *in situ* using *sterilisable* sensors is very limited, however; if one adheres to a rigid definition of *in situ*, perhaps only bare-electrode polarographic and optical methods qualify. The full range of polarographic analyses has never been realised satisfactorily for fermentation, although there is no fundamental reason why this should not eventually be done. Gualandi and Morisi (1966) overcame electrode fouling problems with a relatively massive and expensive sensor in which a silver amalgam electrode was mechanically cleaned by a revolving felt pad. This was designed and used to measure dissolved oxygen, and was very successful; its size, complication and expense, however, militated against use in any except large-scale systems where dissolved oxygen was a critical factor. This probe is not restricted to dissolved oxygen measurement in its potential application, and other solutes could be identified and measured with more versatile electronics which offered a potential sweep and suitable analysis of the resulting electron flow (see Vassos and Ewing, 1972). Failing such applications of polarography, the only other true *in situ* sensors are based on optical properties; that of Harrison and Chance (1970) uses a fluorescence technique to measure NAD–NADH level *inside the cells*, and therefore gives direct information in cell metabolic states.

Apart from such methods, all determinations of chemical composition must be done on samples withdrawn from the culture. Harmes (1971) has performed considerable development work on automatic sampling and analysis, and the coupling of automatic analysers to fermenters has allowed the measurement of almost any variable required. Enzyme electrodes (see Section 5.4.3.4.2) allow a readout to be obtained corresponding to the concentration of the enzyme substrate, although frequent recalibration may be required. Enzymes themselves may be measured in the culture sample, in addition to turbidity and cell shape and size (Wyatt, 1970) and cell ATP content (Hastings, 1968).

Filtrates of the culture (Harmes, 1971) or dialysates (Inagaki, 1965) may be more suitable.

Kinoshita (1972) gives some details of automatic on-line analysis of glucose and glutamic acid. In principle the use of dialysis against a bactericidal solution yields two streams: one of dialysate, which may be chemically analysed for low-molecular-weight species, and the dialysed stream containing stabilised cell suspension and high-molecular-weight species.

5.4.61 SUBSTRATE ANALYSIS

Automatic methods as outlined above may be used to measure mono-, oligo- and polysaccharide constituents. Inagaki (1965) added KCN solution to unfiltered broth and equilibriated by dialysis with ferrocyanide solution. This system, shown in Fig. 5.18, presupposes either sterile conditions right into the automatic analyser, or some buffer capacity between fermenter and analyser, with possible loss of definition of results.

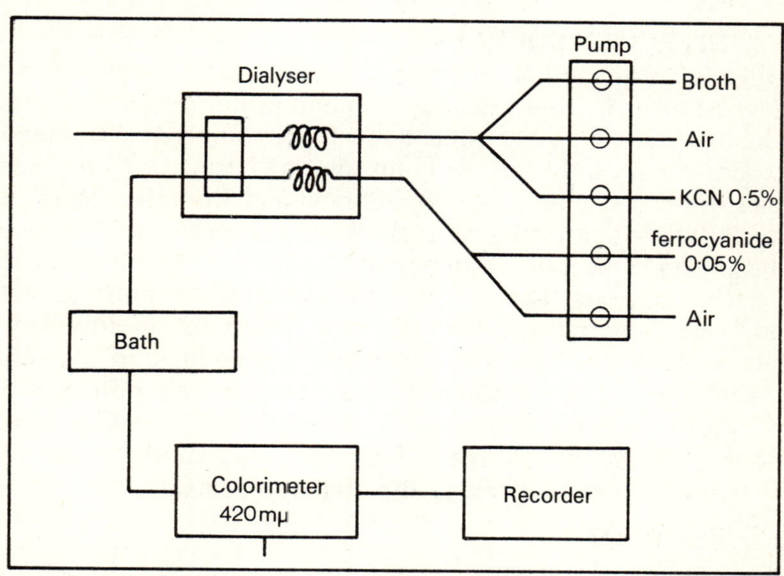

Fig. 5.18
Automatic on-line measurement systems.

Whatever the system employed for analysis, once the sample stream has left the fermenter the problem is one only of automatic analysis, for which an extensive methodology and literature exist; any suitable method of converting substrate concentration into its electrical or pneumatic analogue may be used. For glucose, for example, many methods of determination exist including the 'glucose electrode' (Updike and Hicks, 1967). For other monosaccharides, other enzymes also give specific reactions. For total carbohydrate, several automatic methods exist, of which the best class is probably reaction with sulphuric acid followed by measuring optical density in the U.V. region. For non-carbohydrate substrates, more specific methods must be used. Volatile substrates such as methanol, ethanol, lower paraffins (including those gaseous at fermentation temperatures), give a distinct equilibrium concentration in the gas phase, on which either direct measurement or gas chromotography may be performed. The latter has the advantage of almost perfect specificity, with the determination of several gases and volatiles together. Using suitable detectors, oxygen and carbon dioxide may thus be monitored together with volatile substrates and products.

For non-volatile hydrocarbons, U.V. may be used to some extent, although analysis of extracted discrete samples by h.p.l.c. or gravimetry may be the only absolute methods.

5.4.6.2 NITROGEN SOURCES

No direct *in situ* method is known, except where ammonia is the nitrogen source; in this case the ammonium ion selective electrode may be used. The urea electrode, basically a urease-coated ammonium electrode, could only be used outside the fermenter on discrete samples or on a sample stream (Guilbrand and Montalvo, 1969). For *extracellular* protein nitrogen, analysis may be performed on centrifugates, filtrates or porous-membrane dialysates. Where particulate nitrogen sources are used, a total nitrogen content must be determined, with a correction for cell nitrogen based on turbidity or nucleic acid values. Several assumptions are implicit in the use of such a method. For continuous total nitrogen determinations on whole broth, cell-free broth or on the cell fraction alone, several methods exist based on the Technicon® helical digestion tube, which allows automatic Kjeldahl digestions to be made.

5.4.6.3 PRODUCT

Methods in this group must be as variable as are the properties of fermentation products. Where cell material is the main product, methods are generally straightforward, and will be dealt with in Section 5.4.7.

Volatile products may be measured by gas phase chromatography, as in Section 5.4.6.1, or, in specific cases, by sensitive sensor elements in the gas stream. The 'Pellistor' catalytic element, developed by the Safety in Mines Research Establishment, is used by several companies making proprietary detection systems. Acidic products may be estimated, under conditions of automatic pH control, by measuring the use of alkali, although this is not always proportional to growth (Luedeking and Piret, 1959).

Many products may be determined specifically by automatic analysis on the same sample stream as used for medium constituents (Sections 5.4.6.1 and 5.4.6.2): for example, glutamic acid may be determined by an enzymic system (transaminase or deaminase) using a standard automatic analysis system with a colorimeter wavelength of 555 nm (Inagaki, 1965). Many antibiotics and vitamins may be determined by optical techniques on the sample stream, although two or more wavelengths may be needed to achieve specificity. Enzymes may similarly be determined automatically, although precautions must be taken to ensure full activation, presence of co-enzymes etc. in the analysis.

Fermentations which produce polysaccharides raise the viscosity of the broth. This may be used directly to follow the progress of the fermentation, although viscosity is not generally proportional to concentration (e.g. Kovacs and Kang, 1977). Various integral viscometers exist which could in principle be used in fermenters, or built into sterile loops attached to them, but in large-scale work the contamination risk and engineering complication are minimised by using a sample stream passing through an in-line viscometer cell. Various commercial models exist, and the reviewers have developed a simple, robust and somewhat more economical variant of the in-line viscometer (Priestley and Vincent, in press). This uses a pressure transducer to measure the back-pressure generated by pumping at constant rate (using a Mono pump) through a constriction. After careful calibration this system gave accurate viscosity analogue signals at preset intervals throughout batch production runs, useful in optimisation of the kind detailed in Section 5.2.2. Although used for xanthan gum, the principle is usable for scleroglucan, *Erwinia* polysaccharide, dextran, '*curdlan*' pullulan etc. It should be noted that most of these polysaccharides are very non-Newtonian in rheology, so that apparent viscosity is very shear-dependent. The viscosity/dry weight relationship of gum may vary throughout a batch run, and even through the cell division cycle (Vincent, personal observations), so calibration in absolute units is not generally possible.

5.4.7 Measurement of growth

Various definitions exist for growth. The main alternative, and often conflicting, criteria are increase in cell number and increase in cell dry weight (Mallette, 1969). These different criteria may suit different applications, and will both be considered here in addition to methods depending on measuring other cell properties or single components of cells.

Counting of total microbial cells may be used as an absolute base for other methods, their results then being expressed in cell numbers, although the validity of the assumptions involved is questionable. Electronic and microscopical methods of counting have been compared (Mountney and O'Malley, 1966), and in general electronic methods seem to be at least acceptable in accuracy. Correlation between counts and growth rate is evidently better when using dilution/plating methods, which count only the viable cells. The Coulter Counter technique has great value when used with a pulse height analyser, since cell size distribution is determined in addition to total count (Kubitschev, 1969). This yields information on synchronisation, division, kinetics etc. Total viable counts may be performed; many methods exist, but they mostly have in common a dilution or dispersion of a sample in a non-toxic medium (Postgate, (1969). This is then serially diluted, usually by ten-fold dilutions, to give a final expected population of about 10^2 – 10^3 cells/ml. Postgate describes a technique with dilution steps of $\times 10^4$ claimed to avoid the cumulative errors of serial dilutions.

Dilutions are spotted and spread onto a pre-dried agar medium surface, incubated and counted. An ideal count should have 200–300 colonies per agar plate. Aerobes or non-exacting anaerobes may be counted in this manner or in roll-tube equipment (e.g. Astell). Exacting anaerobes must be grown and counted in shake-tubes or in roll-tubes with pyrogallol-coated plugs (see Hungate (1971) for detailed treatment). The drop-plate method for aerobes on non-exacting anaerobes involves smaller volumes applied to agar in drops (replicate) using laboratory-calibrated Pasteur pipettes (Miles and Misra, 1938) or commercial micropipettes (e.g. Eppendorf).

For an accuracy of estimation within a factor of 2, the Most Probable Number technique may be used. This depends on 'dilution to sterility'. At high dilutions, not all diluted media will grow on incubation, and the original total count may be estimated using statistical tables (see Taylor, 1962).

Many proprietary systems now exist for estimating total microbial count by measuring growth rate (turbidity) in purpose-designed equipment (e.g. A.M.E), although in this case the implicit assumptions about growth rate, adequacy of aeration etc. must be recognised, and the comparability between samples must be relatively poor. Luciferin-based photogeneration for measuring ATP content may also be used for estimating total viable bacteria.

Membrane counts are now used to an increasing extent, particularly for water and air bacteriology (see Mulvany (1969)), but evidently they must be coupled with pre-dilution of samples for fermentation work since populations are generally so high. They have the advantage of allowing total and viable counts on parallel samples.

Total microbial counts may be made by visual counting, in suitable chambers under the microscope (Norris and Powell, 1961), and offer the only absolute method on which *number estimating* methods may be based using turbidity, nephelometry etc. Mallette (1969) discusses such methods in detail in addition to dry weight determination methods, which are used *per se* and also as a calibration basis for optical and chemical methods.

Chemical determinations of nitrogen, phosphorus, carbon, DNA/RNA may also be used, more usefully on separated cells than on whole culture. Dialysis, a standard automatic analysis technique, may be used to remove interfering soluble fractions before performing determinations of the total amount of any constituent in the cellular fraction, although precautions must be taken to avoid growth, cell lysis or retention of extracellular macromolecules if these would interfere with the analysis. Standard methods for discrete samples will not be detailed here. Some automatic or semi-automatic systems are usable on-line or off-line rapidly enough to furnish data useful in following fermentation processes. The Technicon® system allows Kjeldahl-based determination of total nitrogen. The Coleman range (Mazoyer, 1962) uses the Dumas principle, but allows also determination of carbon and hydrogen. Bennett and Williams (1966) reported the phosphorus content of the cell fraction to be a reliable growth index. Determinations of nucleic acid content other than those based on phosphorus content are available, but all methods suffer from interference by nucleotides and nucleosides.

Specific cellular constituents may also be used to measure growth. Diaminopimelic acid is specified for Gram-negative cell wall, and may be used as an index of growth (e.g. el-Shazly and Hungate, 1966), although purely chemical methods must be relatively complicated to offer adequate specificity for automatic analyis and therefore on-line availability of results. In this case, specific enzymes may offer the desired simplification.

It will be evident from this review that often the need for rapid availability of results conflicts with the accurate determination of growth. No case can be made that optical methods are as accurate as dry weight determinations, for instance. Perhaps, for academic studies at least, direct although slower and more time-consuming methods should continue to be used. However, inaccuracies of acceptable level are attainable, and 'indirect' though rapid methods serve best the general aim of growth optimisation.

The following sections present a few examples of parameters which may be measured to give an indirect indication of growth.

5.4.8 Pressure

Solid state pressure transducers have been developed in recent years, capable of producing a pressure analogue using virtually no moving parts. This has led to the use of pressurement for a wide range of applications in which conventional strain gauges were not useable owing to cost or bulk or both (e.g. flow measurement).

One such device is the piezoelectric pressure transducer (Bio-engineering AG, Switzerland). This is supplied in a housing which can be screwed directly into a fermenter port, and includes integral temperature compensation. The signal is transmitted along 4 metres of cable to the amplifier, which generates a 0 – 20 mA signal for a pressure range of 0 – 2 bar.

Another interesting development is the pressure-sensitive transistor or 'Pitran', manufactured by Raytheon. The use of the Pitran is discussed in some detail by Wightman (1973), but the authors are unsure as to whether this device is still available in the U.K.

Apart from use in measuring fermenter headspace pressure and in flow measuring, such transducers may also be used to sense the weight of the contents of a fermenter, and thus form the basis of a turbidostat. When positioned at the bottom of the fermenter (covered by a protective membrane if necessary), the pressure transducer can be used to feed circuitry controlling the opening of a valve when the weight of the fermenter contents reaches a preset level.

5.4.9 Power input:

Various sensors can be used for measuring power supplied to a fermenter, including ammeters in the motor circuit, strain gauges, and dynamometers. Various correlations exist for the evaluation of power supplied by the impeller in terms of the impeller speed (see Vincent and Priestley, 1975), and so it is relevant at this stage to present a brief summary of the methods by which shaft speed may be measured.

D.C. tachometers (with a dynamic range of $> 1000:1$) in their simplest form are essentially d.c. motors used as signal generators, but suffer from problems such as brush wear and electrical noise such as spikes and ripple. Brushless types, in which the magnet rotates and the coil is stationary, offer improved performance but are unreliable at low speeds; it is also necessary to provide external circuitry to sense the position of the magnet and provide the requisite switching.

A.C. tachometers (with a dynamic range of approximately 100:1) can be single- or three-phase generators, with a rectification stage on the output. They suffer from non-linearity at low speeds, however, owing to the voltage drop across the rectifiers. They are adequate for systems with a narrow speed range.

Digital pulse generators are preferable, either in magnetic or optical form. Magnetic systems consist of magnets positioned on the shaft; a magnetic proximity sensor then provides a pulse when the magnet passes through its field of sensing. These systems suffer from irregularities at low speeds. The optical tachometer uses an incremental optical encoder with sophisticated (e.g. microprocessor-based) electronics to count the pulses and convert them to true engineering units. These can be fairly reliable down to low speeds, and have a wide dynamic range of the order of 1000:1.

5.4.10 Viscosity

As was mentioned in Section 5.4.8, the viscosity of polysachharide-forming cultures can be used as an index of product formation provided that the data are interpreted with caution. In addition to the in-line systems such as the Haake VISCONTROL, there are many 'batch' viscometers on the market, many of which were developed for use in the oil industry (see Rein (1978)). The rotational viscometer such as the Brookfield is one such instrument, working on the principle of measuring the viscous drag on a bluff body such as a cylinder or cone rotating in the test fluid. A wide range of shear rates can be obtained by changing the size of the rotating bob and the speed of rotation, enabling rheograms to be constructed with relative ease.

5.5 RECORDING AND DATA LOGGING

5.5.1 Analogue chart recorders

The multichannel chart recorder is the most well-known recording instrument, models being available which can scan up to 36 channels. They are relatively cheap instruments suitable for logging data from systems at steady state or which are changing only slowly; they are unsuitable for observing fast transients (a potentiometric pen recorder should be used for this purpose).

Multichannel recorders operate on a time-sharing basis, in which the signals presented to the instrument are scanned in sequence. The pen, driven either by a servomotor or forming part of the rotor of a linear motor, is then positioned over the chart paper and prints a dot. The frequency with which dots are printed evidently depends upon the number of channels to be scanned and the scanning rate; whether the points printed for each variable are discontinuous or form a continuous line depends upon the scanning rate and the chart speed.

Like digital panel meters, chart recorders usually have a relatively high input impedance (in excess of 1 MΩ), and can therefore in some cases be used to measure directly the output of a transducer. However, in most cases the signal will have to be conditioned in some way in order to use the recorder scale most effectively.

Input ranges to chart recorders may be 0 –10 mV, 0 – 10 mA, 0 – 20 mA, 0 –300 μA. Some instruments (e.g. the Foster Cambridge P12OL) have interchangeable 'function cards' which can be used to change the range of the instrument. Recorders with this facility can be used to directly record such variables as temperature, a.c. volts, frequency, amperes etc. Some recorders also have alarm facilities.

The major disadvantage of chart recorders is their relatively poor resolution; if this criterion is regarded as an essential property of the measuring system, the use of a digital data logger is mandatory.

5.5.2 Digital data loggers

The function of a digital data logger is to record process data in digital form by scanning incoming variables either continuously or at self-initiated intervals. Commercial systems of great sophistication are now available, such as the Leeds & Northrup Trendscan 1000. This system can scan up to 1000 variables at intervals of between one minute to 80 hours, scanning being initiated by means of an integral clock set to either real or process time. A printed record at the conclusion of the scan includes such information as the day, time, value and channel number. An advantage of this particular system is its incorporation of an analogue function, in which data can also be presented in the form of a bar graph with digital printing of the desired significant figures; this is obviously extremely useful for trending or profiling data. The system has been used by the authors to activate remote intermittent sampling devices at preset intervals as part of the scanning sequence.

The input to this system is normally 40 mV FSD, and data are recorded in mV. Custom-designed software is available, however, to enable the print record to be in terms of true engineering units. Additional output stages are available to record data on punch cards, magnetic tapes etc., either locally or via the telephone system.

REFERENCES

Bennett, E. O. and Williams, E. P. (1957), *Appl. Microbiol.* **5**, 14–16.

Briggs, R. (1971), Institute of Measurement and Control Symposium on *Measurement and Control of Industrial Pollution.*

Brown, D. E. (1970), In *Methods in Microbiology,* Vol. 2, pp. 125–174, ed. Norris, J. R. and Ribbons, D. W., Academic Press, London and New York.

Driscoll, F. F. and Coughlin, R. F. (1975), *Solid State Devices and Applications,* Prentice-Hall, New Jersey.

el-Shazly, K. and Hungate, R. E. (1966), *Appl. Microbiol.,* **14**, 27–30.

Epps, H. M. and Gale, E. F. (1942), *Biochem. J.,* **36**, 619.

Barford, J. P. and Hall, R. J. (1978), *Process Biochem.* **8**, 22–29.

Chain, E. R. and Gualandi, G. (1954), *Rend. ist sup Sanita,* **17**, 5.

Gale, E. F. (1940), *Bact. Rev.* **4**, 135–176.

Gualandi, G. and Morisi, G. (1966), *Biotech. Bioengng.* **8**, 621.

Guilband, G. G. and Montalvo, J. G. (1969), *J A C S,* **91**, 2164.

Harmes, C. S. (1971), S.I.M. Annual Meeting Paper.

Harrison, D. E. F. and Chance, B. (1970), *Appl. Microbiol.* **19**, 446.

Hastings, J. W. (1968), *Ann. Rev. Biochem.* **37**, 597.

Herbert, D., Elsworth, R. and Telling, R. C. (1956), *J. Gen. Microbiol.* **14**, 601.

Hnatek, E. R. (1975), *Applications of Linear Integrated Circuits,* Wiley, New York.

Hospodka, J. (1966), *Biotech. Bioengng.* **8**, 117.

Humphrey, A. E. (1974), *Chem. Engng.* Dec. 9, 99.

Hungate, R. E. (1971), In *Methods in Microbiology,* Vol. 3b, ed. Norris, J. R. and Ribbons, D. W., Academic Press, London and New York.

Inagaki, T. (1965), *J. Soc. Instr. Autom. Control (Japan)* **4**, 545.

Jacob, H. E. (1970). In *Methods in Microbiology,* Vol. 2, p. 91, ed. Norris, J. R. and Ribbons, D. W., Academic Press, London and New York.

Kinoshita, K. (1972). In *The Microbial Production of Amino Acids,* Kodansha, Tokyo, Wiley, New York.

Kubitschev, H. E. (1969). In *Methods in Microbiology,* Vol. 1, ed. Norris, J. R. and Ribbons, D. W., Academic Press, London and New York.

Kovacs and King (1977).

Lengyel, Z. L. and Nyiri, L. (1965), *Biotech. Bioengng.* **7**, 91.

Lineweaver, H. and Burke, D. (1934). *J. Am. Chem. Soc.* **56**, 658.

Luedeking, R. and Piret, E. L. (1959), *J. Biochem Microbiol. Tech. Engng.* **1**, 393.

MacLennan, D. G. (1970). In *Methods in Microbiology,* Vol. 2, p. 1, ed. by Norris, J. R. and Ribbons, D. W., Academic Press, London and New York.

Mallette, M. F. (1969). In *Methods in Microbiology,* Vol. 1, ed. Norris, J. R. and Ribbons, D. W., Academic Press, London and New York.

Mazoyer, R. (1964), *Bull. Ass. franc. Etude Sol.* 282.

Michaelis, L. and Menten, M. L. (1913), *Biochem. Z.* **49**, 333–369.

Miles, A. A. and Misra, S. S. (1938). *J. Hyg. Camb.* **38**, 732.

Monod, J. (1950), *Ann. Inst. Pasteur (Paris)* **79**, 390.

Moraine, R. A. and Rogovin, P. (1971), *Biotech. Bioengng.* **13**, 381.

Moraine, R. A. and Rogovin, P. (1973), *Biotech. Bioengng.* **15**, 225.

Mountney, G. J. and O'Malley, J. (1966), *Appl. Microbiol.* **14**, 845.

Mulvany, A. A. (1969). In *Methods in Microbiology,* Vol. 1, ed. Norris, J. R. and Ribbons, D. W., Academic Press, London and New York.

Munro, A. L. S. (1970). In *Methods in Microbiology,* Vol. 2, p. 39, ed. Norris, J. R. and Ribbons, D. W., Academic Press, London and New York.

Munson, R. J. (1970). In *Methods in Microbiology,* Vol. 2, p. 346, ed. Norris, J. R. and Ribbons, D. W., Academic Press, London and New York.

Nagai, S., Nishizawa, Y. and Aiba, S. (1971), *J. Gen Microbiol.* **59**, 163.

Nilsson, H. and Mosbach, K. (1978), *Biotech. Bioengng.* **20**, 527.

Nyiri, L. (19PP), *Advances in Biochemical Engineering,* Vol. 2, p. 49.

Norris, K. P. and Powell, E. O. (1961), *J. Roy. Microscop. Soc.* **80**, 106.

Ohashi, M. J. (1958), *J. Chem. Soc. Japan* **61**, 1001.

Patching, J. W. and Rose, A. H. (1970). In *Methods in Microbiology* Vol. 2, p. 23, ed. Norris, J. R. and Ribbons, D. W., Academic Press, London and New York.

Peringer, P. (1972), *Biotech. Bioengng.* **14**, 411.

Pirt, S. J. (1969), *Microbial Growth,* Cambridge University Press, pp. 199–222.

Postgate, J. R. (1969). In *Methods in Microbiology,* Vol. 1, ed. Norris, J. R. and Ribbons, D. W., Academic Press, London and New York.

Rabotnowa, I. (1963), *Die Bedeutung physikalisch-chemisch Faktoren (pH u. rH_2) fur die Lebenstatigkeiten der Mikroorganismen,* VEB Gustav Fischer Verlag, Jena, German Democratic Republic.

Rein, S. W. (1978), *Lubrication,* **64**(2), 13.

Steel, R. and Maxon, W. D. (1962), *Biotech. Bioengng.* **4**, 231.

Squires, R. W. and Hosler, P. (1958), *Ind. Engng. Chem.* **50**, 1263.
Tengerdy, R. P. (1961), *J. Biochem. Microbiol. Tech. Engng.* **3**, 255.
Updike, S. J. and Hicks, G. P. (1967), *Nature,* **214**, 986.
Vassos, B. H. and Ewing, G. W. (1972), *Analog and Digital Electronics for Scientists,* Wiley-Interscience, New York.
Vincent, W. A. (1974a), *Process Biochem.* **9**(3), 19.
Vincent, W. A. (1974b), *Process Biochem.* **9**(7), 30.
Vincent, M. A. and Priestley, G. (1975). In *Handbook of Enzyme Biotechnology,* p. 163, ed. Wiseman, A., Ellis Horwood, Chichester.
Vincent, W. A. (1972). Lectures on Process Variables in Enzyme and Fermentation Biotechnology, University of Surrey, Guildford.
Wightman, E. J. (1972), *Instrumentation in Process Control,* Butterworth, London.
Wimpenny, J. W. T. (1969). In *Microbial Growth,* Cambridge University Press.
Wyatt, P. J. (1970), *Nature,* **226**, 277.

Chapter 6

Developments in the Immobilisation of Microbial Cells and their Applications

Dr. Peter S. J. Cheetham, Tate and Lyle Ltd., The University Reading

NOMENCLATURE AND ABBREVIATIONS

ATP	Adenosine triphosphate	
BOD	Biological oxygen demand	
CMP	Cytosine monophosphate	
EDTA	Ethylenediaminetetra–acetic acid	
DEAE	Diethylaminoethyl–	
a	Area of viable cells per unit volume of microbiol mass	L^{-1}
a_m	External surface area of a model micro-organism	L^2
a_R	The external surface area of a single representative cell	L^2
a_p	The external surface area of the permease zone	L^2
D_e	The effective diffusion coefficient	L^2T^{-1}
D_m	The molecular diffusivity of a substrate	L^2T^{-1}
S	Bulk substrate concentration	ML^{-3}
S'	The substrate concentration external to the permease zone	ML^{-3}
S''	The substrate concentration in the metabolic zone	ML^{-3}
K_m	Michaelis – Menten constant	ML^{-3}
K'_m	Effective Michaelis – Menten constant	ML^{-3}
L	The thickness of an immobilised cell pellet	L
V_{max}	The maximum rate of reaction	T^{-1}
V'_{max}	The effective maximum rate of reaction	T^{-1}
V_R	The volume of a single micro-organism	L^3
α	Rate coefficient	$ML^{-2}T^{-1}$
α_R	Rate coefficient for metabolic activity	$ML^{-3}T^{-1}$
α_p	Rate coefficient for permease transport	ML^{-3}
β	Rate equation coefficient	ML^{-3}
β_R	Rate equation coefficient for permease transport	ML^{-3}
β_p	Rate equation coefficient for permease transport	ML^{-3}
ϵ	Porosity of an immobilisation support material	–
η	Effectiveness factor	–
η'	Effectiveness factor in the presence of inhibitor	–
τ	Tortuosity of a pore	–
υ	Initial rate of reaction	$ML^{-3}T^{-1}$
υ'	Initial rate of reaction in the presence of inhibitor	$ML^{-3}T^{-1}$

6.1 INTRODUCTION

Micro-organisms have been used in the production of foodstuffs for at least 4000 years, but now seven main, distinct types of biochemical catalyst may be distinguished. These are actively growing whole cells, non-growing whole cells or spores, modified cells such as cell debris, crude or purified soluble enzymes, insolubilised enzymes, soluble immobilised enzymes and immobilised viable or non-viable whole cells.

Comparatively little work has been carried out on the immobilisation of cells compared with the voluminous literature on fermentations and immobilised enzymes, but the potential of immobilised cells is just beginning to be recognised as shown by the work described at the 1st European Congress on Biotechnology held at Interlaken in September, 1978, and a flood of publications is expected in the near future. In this 'state of the art' review the academic and patent literature has been surveyed up to December, 1978, in an effort to describe existing work, to elucidate the characteristics and present uses, and to speculate about the potential of immobilised whole cells. However, I suspect that some of the most interesting industrial work dealing with immobilised cells has not yet been made fully public. In the existing literature important information concerning the biochemical, physiological, morphological and genetic states of the cells and comparisons of the activities and stabilities of free and immobilised cells, are very often not given. Thus fully comprehensive descriptions of the various methods of cell immobilisation and types of immobilised cell preparations, and their uses, cannot be made. A comparison of immobilised cells with other types of industrially useful catalysts is made, and the current industrial applications of immobilised cells are comprehensively described in an attempt to show how immobilised cells could play a much more important role in the biochemical industries. In particular, our experience using microbial cells immobilised in calcium alginate gel pellets, and the kinetics of such gel entrapped cells, are considered in some detail. Emphasis has also been placed upon the interdisciplinary nature of both the pure and applied research needed to invent, develop and apply immobilised cell processes.

For the purpose of this review cell immobilisation is defined as any technique which severely limits the free diffusion of cells so enabling their easy recovery, usually by centrifugation or filtration, but also on occasions by gravitational sedimentation or magnetic attraction. Diffusion can be restricted by aggregating the cells or by containing the cells in, or attaching them to, a solid support, so enabling the cells to be used continuously or to be reused in batch processes. This is a great economic advantage as the cost of the biological catalyst is usually a major factor determining the cost of the final product (Fig. 6.1). Immobilisation often enhances the operational

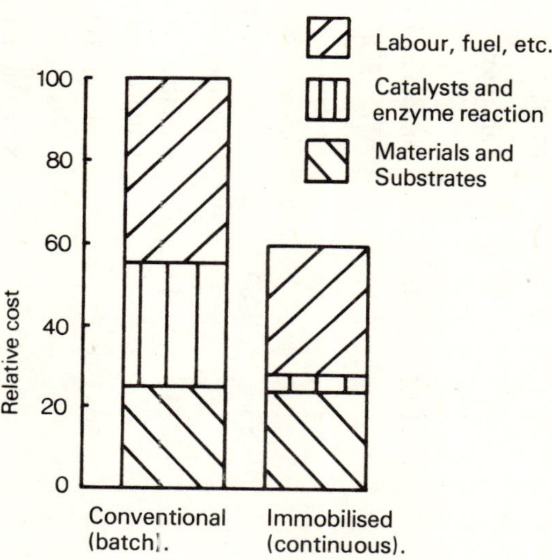

Fig. 6.1 – A comparison of the relative costs of the industrial production of L-aspartic acid using intact or immobilised *E. coli*. This figure was first published by Chibata and Tosa (1977).

stabilities of the enzyme activities of the immobilised cells (Table 6.1), and greatly decreases the size of reactor required to achieve a given productivity. This definition of cell immobilisation includes naturally aggregated or immobilised cells (Long, 1977), immobilised cell debris, or cells lysed after immobilisation (Chibata and Tosa, 1977), but excludes cells retained for use either in a fermenter or a second vessel and the natural cell films (Newman, 1974) which are used in sewage treatment (Larson and Dimmick, 1964) or traditional vinegar manufacture (Abson and Todhunter, 1967). Often when only a single enzyme of the immobilised cell is used or when the cell has lysed, the biocatalyst is referred to as an immobilised enzyme. Also when cell division occurs within the immobilised cell preparation or when enzymes are immobilised onto the surface of cells the distinction between immobilised enzyme, immobilised cell, and fermentation processes can be blurred.

At present immobilised whole cells have academic interest, especially as models of membrane-bound enzymes. Medical and analytical applications would appear to be limited by the non-specificity of

Table 6.1

A comparison of the activities and stabilities of various immobilised cell preparations

Product	Micro-organism	Half-life (days)⁺	Half-life of free cells (days)	μ moles of product/h/g wet cells	μ moles of product/h/g g gell	Reference
Aspartic acid*	*E. coli*	120(37°C,pH8.5)	11	12,200	1,220	Chibata *et al* (1974e); Tosa *et al* (1974)
Malic acid*	*Brevibacterium ammoniagenes*	52.5 (37°C,pH7.5	6	14.3–16.5	2.0–2.3	Yamamoto *et al* (1976, 1977)
Citrulline	*Pseudomonas putida*	140 (37°C,pH6.0)	–	371	51.5	Yamamoto *et al* (1974a)
6-aminopenicillic acid	*E. coli*	42 (30°C,pH8.5)	–	112.5	45	Chibata *et al* (1975); Sato *et al* (1976)
Urocanic acid	*Achromobacter liquidum*	180 (37°C,pH9.0)	–	157	21.8	Yamamoto *et al* (1976)
Prednisolone*	*Corynebacterium simplex*	– (25°C,pH7.0)	–	83	69	Larsson *et al* (1976), Ohlson *et al* (1978)
Novel disaccharide	–	166 (30°C,pH7.0)	1.75	234	47	Cheetham *et al*, unpublished patent
Glucose and Fructose	*Actinomyces missouriensis*	45 (30°C, pH7.5)	–	50–350	–	Linko *et al* (1977)
Ethanol	*Saccharomyces cerevisiae*	10 (25°C)	–	–	–	Kierstan and Bucke (1977)
Ethanol	*Kluyveromyces marxianus*	15 (25°C)	–	–	–	Kierstan and Bucke (1977)

* In industial use.
\+ The half-life is usually the time taken to lose half the original activity.

immobilised cells and the diffusion barrier presented to a potential substrate by the cell wall and membrane, except in specialised uses such as BOD measurements. For example, Karube *et al,* (1977a,b) have devised a simple; cheap, quick, reproducible and stable method of measuring BOD, by entrapping soil micro-organisms in collagen around an oxygen electrode, or by using *Clostridium butyricium* entrapped in polyacrylamide gel around a platinum electrode. Similarly, an L-glutamate sensing electrode has been made by entrapping cells at the surface of a potentiometric ammonia analyser (Rechnitz *et al,* 1978; Fig. 6.2). Suzuki and Karube (1978) have also described a series of immobilised cell electrodes used to detect a variety of biochemicals such as cephalosporins. Such electrodes may be the forerunners of a wide range of biochemically based analytical devices.

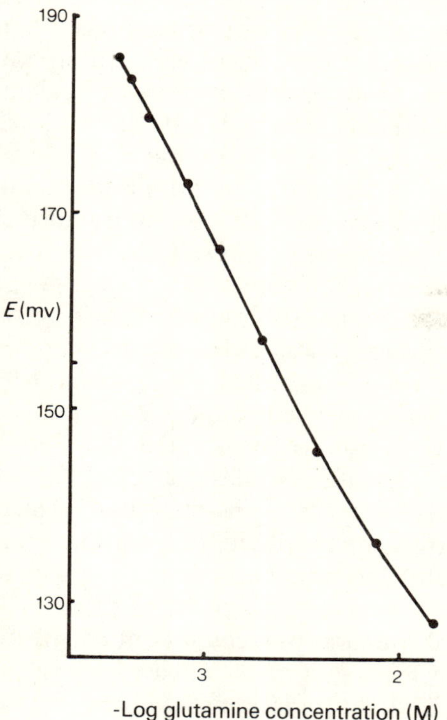

Fig. 6.2 – The response of a bacterial sensor to glutamine. Measurements of glutamine were made in reconstituted control serum diluted 1 : 5, (Rechnitz *et al,* 1978).

In industry the formation of a variety of antibiotics, steroids, coenzymes and organic acids, including amino acids, using immobilised cells have already been described (Chibata and Tosa, 1977). Applications include the use of viable, non-viable and disrupted cells. Often conversions requiring only a single enzyme can be operated using non-viable cells, for example, some of the glucose isomerase processes (Bucke, (1977); whereas multi-step transformations which utilise much of the cells' complete metabolism usually require intact, viable cells. Examples are the fermentation of inulin and glucose to ethanol using alginate immobilised *Klyveromyces marxianus* and *Saccharomyces cerevisiae* respectively (Kierstan and Bucke, 1977), or the conversion of glucose to hydrogen in a biochemical fuel cell in which *Clostridium butyricium* entrapped in polyacrylamide gel is used to analyse the organic materials in waste waters (Karube *et al,* 1977c). Furthermore protozoa, organelles and microbial cells are of comparable complexity and so it should be possible to immobilise them all by very similar methods. For instance, Kierstan and Bucke (1977) immobilised yeast cells, chloroplasts from spinach, and mitochondria from avocado pears in calcium alginate gels, which produced ethanol, carried out the Hill reaction, and produced ATP respectively. Other examples of immobilised organelles are the rat liver mitochondria immobilised to alkyl silyated glass beads by Arkles and Briniger (1975), the use of immobilised chloroplasts to produce hydrogen (Packer, 1976) and the chromatophores of *Rhodospirillum rubrum* which are entrapped in polyacrylamide with 40% retention of photophosphorylating activity. The stability of these chromatophores was enhanced by immobilisation but the reaction was diffusion controlled (Yang *et al,* 1976). Microbodies from *Kloeckera* have been entrapped in albumin cross-linked with glutaraldehyde and photo-cross-linked in polyethylene glycol hydroxyethylene acrylate. Catalase, alcohol dehydrogenase and D-amino acid oxidase activity were measured (Tanaka *et al,* (1978). Similarly, composite immobilised preparations containing different cell types, or cells and enzymes, or cells and organelles, could be used to combine the advantages of these different biocatalysts. The most complicated preparations could contain the many different cell types responsible for digestion of sewage, microbial leaching of low grade ore, or pesticide degradation (Senior *et al,* 1976).

6.2 A COMPARISON OF IMMOBILISED CELLS WITH ALTERNATIVE CATALYSTS

No single form of catalyst can be expected to be universally applicable and successful. The ideal should be a versatile, fully understood range of industrially suitable biological catalysts, one or more of which can be applied easily to any particular task. With a better understanding of cell immobilisation techniques and a greater appreciation of the advantages of using immobilised cells, then many more useful practical applications of immobilised cells should be expected.

6.2.1 Comparison with chemical catalysts

When compared with chemical catalysts immobilised cells have a number of advantages. These include the variety of reactions catalysed, the mild conditions employed, and thus the low amounts of energy employed; the high degrees of conversion and the specific reactions obtained; and the smaller amounts of pollutants produced. In particular, many organic chemists are attracted to enzyme based catalysts because of the abilities of the enzyme to distinguish between enantiomers and to distinguish enantiotropic groups and the faces of molecules possessing prochiral centers (Jones, 1976a,b), and because such specific reactions can be carried out in aqueous solvents. The speed and specificity of biochemical reactions may also be of use in the preparation of radiochemicals containing short-lived isotopes. However, biological catalysts, such as immobilised cells, often have complex requirements such as co-factors and are usually more labile than chemical catalysts, so requiring precise control of the reaction conditions. Economic factors also influence the choice of catalyst; for instance, chemical processes utilise large amounts of energy and are often based on petrochemicals, whereas biological catalysts do not require large energy inputs and often use renewable biological resources. Thus the present-day use of biological catalysts, such as immobilised cells, in the food and pharmaceutical industries may be followed by their use in the heavy chemical industries in the future.

6.2.2 Comparison with fermentation

At present fermentation is the dominant form of industrial biological catalyst, its great advantage being its versatility, as illustrated by the wide range and sophistication of the processes in use at the present time. However, complex and often expensive nutrients are required, and during fermentation much of these nutrients are used as energy sources, even if the required biochemical is formed in the stationary phase after cell multiplication has been completed. Continuous operation may also be difficult, and capital and running costs are expensive, especially when high rates of aeration and sterile conditions are required. Furthermore, after the fermentation has finished isolation of the desired product is often difficult and expensive, and the cell mass and exhausted fermentation medium must be disposed of.

By comparison immobilised micro-organisms are convenient to handle, appear to be less susceptible to microbial contamination, and permit easy separation of product from the catalyst. The use of immobilised cells enables greater control throughout the reaction, which should ultimately improve the yield and quality of the product. Because the stabilities of cells are usually increased by immobilisation (Table 6.1) continuous use or re-use in batch operations is made easier. Furthermore, because non-dividing immobilised cells require only maintenance energy, yields of product will be greater than in fermentation methods.

Venkatasubramanian *et al,* (1978) have carried out an economic comparison between the batch production of the flavouring agent monosodium glutamate by fermentation, and continuously using collagen immobilised *Corynebacterium* cells. The total fixed capital investment, operating and overhead costs were lower for the immobilised cell plant, but overall production costs were higher because of the need to replace the immobilised cells at intervals. However, because of the lower investment required by the immobilised cell process, this method resulted in a 50% return on capital invested compared to a the 36% return obtained when using the fermentation process.

6.2.3 A comparison with free cells or free enzymes

Free cells or enzymes are very difficult to re-use or to use continuously as both catalysts are too small to filter, and recovery by centrifugation is too expensive. Immobilised cells are often more stable than the equivalent free cells or enzyme, while immobilised cell processes are easier to automate and enable the advantages of various reactor configurations to be exploited. These include rapid pH and temperature control, good gas transfer in stirred reactions, and the minimisation of product inhibition in packed-bed reactors. Also as cells are used more efficiently, fewer cells have to be grown and so less fermentation wastes have to be disposed of. However, the cost of immobilisation and any loss of activity during immobilisation must be borne. Furthermore, the immobilised catalyst will always occupy a larger reactor volume than the free enzyme, so increasing capital and running costs.

6.2.4 Comparison with immobilised enzymes

The use of immobilised cells should be preferred to immobilised enzymes for processes requiring the use of coupled enzymes, expensive co-factors and energy sources such as NADH or ATP; the use of complete metabolic pathways or the complete metabolism of the whole cells. Thus degradative reactions and synthetic reactions which require energy may be efficiently coupled so making immobilised cell technology especially useful for biosynthetic processes. Examples already explored include the formation of CDP-choline from CMP and choline or ethanolamine by immobilised yeast cells (Kimura *et al,* 1978), and the continuous synthesis of coenzyme A using 5 enzymes of immobilised *Brevibacterium ammoniagenes* (Shimizu *et al,* 1975). The choice between the use of immobilised enzymes or immobilised cells is similar in many respects to the more familiar choice between the use of purified or crude soluble enzyme as biocatalyst. This is because the immobilised cell and crude enzyme are cheaper, and larger quantities are available, but they are less specific catalysts than immobilised enzymes or the purified soluble enzymes.

The alternative to the use of immobilised cells is to use increasingly sophisticated immobilised enzyme preparations utilising immobilised co-factors (Mansson *et al,* 1976), co-immobilised enzymes (Srere *et al,* 1973; Lawrence and Okay, 1973; Tramper *et al,* 1978), complex enzyme fractions, such as in the synthesis of Gramicidin S (Hamilton *et al,* 1974), and enzymically active proteins formed by solid-phase chemical synthesis. For instance, Cousineau and Chang (1977) have combined two of these features using an encapsulated multi-enzyme system involving *in situ* co-factor regeneration. To ensure efficient substrate transfer biocatalytic complexes have to be reconstituted using the correct molar amounts of individually purified enzymes and energy and co-factor regenerating systems orientated in the correct spatial arrangements.

Using immobilised cells, enzyme-cell or enzyme-cell debris separation is avoided, and the expensive, lengthy, and tedious extraction and purification of enzymes, often in low yields, and by procedures which are difficult to scale-up, are not required. Furthermore, there is no loss of activity during immobilisation because the enzyme is maintained in its correct native environment inside the cells and so is not subject to stresses such as covalent modification, or conformational deformation which often occur during immobilisation of enzymes. The cell membrane also protects the enzyme from mechanical denaturants such as shear forces and gas bubbles, and excludes many chemical denaturants such as heavy metals and organic solvents. In most immobilised cell preparations the cells are made inaccessible to potentially invasive micro-organisms, and nitrogen free substrates are used, so maintenance of sterile operating conditions is not required. Thus the operational stabilities of immobilised cell enzyme activities are often greater than those of the corresponding immobilised enzyme, particularly if the immobilised cell activities can be regenerated *in situ.*

By immobilising cells, enzymes which are located in the cytosol or bound either to the inside or outside surfaces of the cell membrane can be immobilised by the same technique. Membrane-bound enzymes should be especially suited to cell immobilisation as their activity and stability often depend on continued interaction with the cell membrane. Many cells which naturally grow in contact with biological or non-biological surfaces, such as skin or soil, may be more active and stable after immobilisation, and immobilisation can be used to keep high-floc cells permanently dispersed. Other excellent candidates for cell immobilisation are the mixed function oxygenases, because of their instability, the complexity of their reaction mechanisms and their co-factor requirements which would otherwise necessitate the co-immobilisation of electron transport proteins in the correct relative orientations to the oxygenases. There is also an absolute need for cell immobilisation when an enzyme cannot be extracted in a stable form, is easily inactivated during immobilisation, or when the regulating mechanisms associated with the metabolic pathways are required. It is obvious that cell immobilisation cannot be used for processes involving extracellular enzymes unless it is acceptable for the enzymes to continuously leak out of the immobilised cells during use, such as the α-amylase produced by *Bacillus subtilis* entrapped in polyacrylamide (Kokuba *et al*, 1978). In batch use the α-amylase formed was three times that formed by washed cells, with the enzyme produced increasing with re-use until a steady state was reached after 7 cycles.

Immobilised cells can be considered to be in an artificial stationary phase induced by nutrient deficiency or by lack of space available for new cells, so that cell growth and product formation are not associated. Thus the synthetic activity of the cells can only be maintained for a limited period, depending on the stabilities of the relevant enzymes. Decay can be due to reversible or irreversible inactivation of the enzymes, loss of intermediary metabolites or inhibition by metabolites. Because of the lack of nutrients only non-growth-associated metabolic functions are carried out by the cell, such as osmotic and pH regulation and the turnover of macromolecules. This maintenance energy requirement varies with the type of cell and the environmental conditions. It appears that prolonged cell viability is associated with a low rate of endogenous metabolism closely matched to maintenance energy requirements (Thomas and Bath, 1969), which may explain the great stability of most immobilised cell activities. The operational stability of immobilised cells may often be greatly enhanced by regeneration of the enzyme activities of the immobilised cells. Regeneration may be achieved either by re-induction, or by causing the immobilised cells to divide *in situ*.

Cell division will usually be limited by the space available inside the gel, and by the availability of nutrients. Thus, during operation using nitrogen free substrates cell division can usually be prevented until the low activity of the immobilised cells indicates that regeneration is desirable. Otherwise a proportion of the newly formed cells will leak out of the support material and contaminate the product stream during operation. Ability to achieve regeneration indicates that the immobilisation technique used is very mild and that whole metabolic pathways of the immobilised cells can be utilised. Examples of *de novo* formation of immobilised cells are the 3-ketosteroid Δ^1 dehydrogenase of *Curvularia lunata* Larsson *et al,* 1976; Ohlson *et al,* 1978) and the benzene degrading activity of *Pseudomonas putida*, of which 40–70% was lost on immobilisation in polyacrylamide but which could be restored by incubation in media containing benzene and succinate (Sommerville *et al,* 1976). More recently the same group have demonstrated similar effects using immobilised *Pseudomonas aeruginosa* to oxidise benzoic acid to catechol (Sommerville and Mason, 1978).

An enzyme engineering approach to cell immobilisation is illustrated by our work on cell entrapment in calcium alginate gels, the philosophy being that the investigation of physical effects, such as substrate transfer rates and the mechanical stability of the immobilised cells when used in reactors, should precede biochemical studies. Firstly studies on the support material were carried out to determine its mechanical properties, in particular its tendency to abrade when used in stirred reactors and its resistance to compression and channelling in packed-bed reactors. Such studies helped to determine the most useful reactor configuration and its maximum size, packed-bed reactors being a natural first choice as continuous operation is most easily achieved. The mass-transfer properties of the support were also examined as the porosity of the support to a proposed substrate may be of critical importance in determining the activity of the final immobilised cell preparation. Similarly, a cell immobilised on a charged support would be favoured kinetically if the substrate is of opposite charge.

Biochemical studies can be carried out on a variety of systems of varying complexity. The simplest possible systems require only a single enzyme activity, and so non-viable cells can be used. Determination of the pH, temperature and substrate optima for the conversion, and effectiveness factors for the immobilised cells should be carried out, followed by optimisation of the operational stability of the immobilised cells. Experimentation can then graduate to the use of viable cells in which the enzyme activity requires *in vitro* co-factor regeneration, or a number of enzyme activities acting sequentially are used.

Ultimately, complete metabolic pathways or the complete metabolism of the whole cell can be used. Immobilisation of cell debris and subcellular organelles may also be studied in the same way.

6.2.5 Disadvantages of immobilised cells

Immobilised whole cells are likely to produce less pure products than immobilised enzymes, because unwanted side reactions and further metabolism of the desired biochemical may occur. Therefore, screens of potentially useful micro-organisms for cell immobilisation should include a survey of side reactions. These undesired reactions can often be eliminated by selective denaturation or inhibition of the unwanted enzyme activities of the cell, or by use of substrates of greater purity. For instance, formation of the side product succinic acid during malic acid production can be prevented by the addition of bile salts to the substrate solution (Sato *et al,* 1976). The role of the bile salts may be to selectively denature the enzymes required for succinic acid formation or more probably to impair cell wall integrity, so allowing loss of the co-factors required for succinic acid formation. This latter hypothesis is especially likely, since after bile salt treatment the activity of the immobilised cells was enhanced, presumably owing to the loss of the diffusion barrier to substrate and product transfer caused by an intact cell membrane. Further metabolism of the desired product can also be prevented. For instance, in the production of urocanic acid the enzyme urocanase is selectively heat denatured at 70°C for 30 min. (Yamamato *et al,* 1974b). Alternatively, selection of cells possessing a large ratio of synthetic to degradative activity can be used, such as in the synthesis of 6-amino penicillic acid, where the penicillin amidase activity of immobilised *Escherichia coli* cells was 10 times greater than the penicillinase activity (Chibata *et al,* 1977). Selection of the best micro-organism is also important; for instance, in the formation of L-citrulline most organisms converted the citrulline into L-ornithine using ornithine transcarbamylase, but Yamamato *et al,* (1974a) discovered that *Pseudomonas putida* did not further metabolise the desired product. Using immobilised cells, the yield of product will always be lower than that obtaining with the corresponding immobilised enzyme because of the substrate consumed to provide maintenance energy for the cells.

Contamination of the final product by cells or materials derived from the immobilised cells may also occur, particularly as cell lysis has been frequently observed when cells have been used for long periods. The cells lost may be either the originally immobilised cells, or more usually cells produced by multiplication of the immobilised cells (Cheetham *et al,* 1979; White and Portno, 1978). Only the originally immobilised cells can be leached out of the column when the reactor is operated in the absence of antibiotics and in the presence of antibiotics. However, cell leakage is likely to be especially marked when the complete metabolism of the cell is retained after immobilisation so that division of the immobilised cells is possible.

Because of the relatively large size of immobilised cells, high molecular weight substrates can only be metabolised by immobilised cells used in ultrafiltration cells, the low molecular weight products passing through the ultrafiltration membrane. Additional disadvantages are that the activity of the immobilised cells per unit reactor volume will always be lower than that of the corresponding immobilised enzyme, and so larger reactors or longer residence times will be required to achieve the same productivity. The catalyst site density of immobilised cell preparations will be lower than in the corresponding immobilised enzyme preparations, and so competition between simultaneous enzymic and non-enzymic reactions will be enhanced. An example of such a non-enzymic reaction is psicose formation during glucose isomerisation (Bucke, 1977). Diffusional limitations (mass-transfer effects) on reaction rates using immobilised cells will be large because of the additional diffusional barrier presented by the cell wall and membrane, so emphasising the need for highly porous immobilisation supports. Finally, the activity and stability of immobilised cell preparations may be adversely affected by the action of intracellular proteases, necessitating the use of protease inhibitors such as aminocaproic acid.

6.3 A DESCRIPTION OF IMMOBILISED CELLS

Microbial cells rather than those derived from plant or animal sources are usually used because a constant supply, large quantities, and even quality can usually be guaranteed and because microbial enzymes are usually more stable. Genetic and environmental manipulation to increase yields of cells (Demain, 1971), the enzyme activity of the cells, or to produce altered enzymes (Betz *et al*, 1974) may be employed, these techniques being especially important for the activity or stability limiting enzyme in multi-enzyme conversions.

Many of the physical and chemical methods developed for enzyme immobilisation have also been used for cell immobilisation. These techniques can be broadly described as the formation of cell aggregates, entrapment and encapsulation by inert, usually gellified or porous semi-permeable materials, adsorption to charged supports such as ion-exchange resins, and co-polymerisation with, or covalent binding to, chemically activated supports. The covalent and adsorption immobilisation techniques enable the chemical heterogeneity of the cell surface to be exploited.

Immobilisation supports may be inorganic or organic materials and should be inert to chemical, physical and biological degradation. The support particles may vary in size, shape, density and porosity and can be in the form of particles, sheets, tubes, fibres or cylinders. The support particles may vary in size, shape, density charge and porosity and can be in the form of particles, sheets, tubes, fibres or cylinders. The support should be capable of regeneration or be sufficiently pressure drops and compression problems in packed-bed column reactors and should be sufficiently non-friable to withstand continuous, lengthy use in stirred-tank reactors. Resistance to gases generated by the immobilised cells, exposure to organic solvents, and good operational and storage stability over a wide range of conditions are also an advantage.

In any instance the immobilisation technique used will depend on the scientific, engineering and economic aspects of the process. A cell immobilisation method should be simple and reproducible so as to reduce the loss of activity upon immobilisation, and should avoid the use of denaturing conditions such as extremes of pH or temperature, and the use of organic solvents. The method should also be cheap, generate no fines, and be capable of use over a wide range of conditions so as to enable use of a wide variety of different cells and the co-immobilisation of different types of cells. The immobilisation method of choice should allow easy control of the amount of cells immobilised, and cells should not leak from the support during storage or operational use. During immobilisation, use of dangerous apparatus and toxic and corrosive chemicals should be avoided. For instance, activation of cellulose based supports with cyanogen bromide or 2-amino 4,6-dichloro-S-triazine and use of many protein cross-linking agents, are hazardous operations, especially on a large scale. So as to prevent undesirable partitioning effects between support, substrate, and product, the support material should be uncharged and hydrophilic. The diffusional restriction on reaction rates can be minimised by even dispersion of the cells in the immobilisation support, which is facilitated by the use of low-flocculating cells and small highly porous support particles.

During the harvesting and immobilisation of cells maximum enzyme activities can be most easily retained by minimising pH, osmotic pressure, and temperature shocks. During use, microbiological contamination can be avoided by prefiltration of the substrate or by sterilisation of the substrate with ultra-violet light or by pH or temperature shocks.

6.4 METHODS OF IMMOBILISING WHOLE CELLS

Most of the techniques used for enzyme immobilisation have also been used for cell immobilisation, the chief difference being the larger size of the microbial cells. Unfortunately, but understandably, most work has been oriented towards formation of satisfactory immobilised cell preparations on a laboratory scale, and little systematic, comparative, methodological or large-scale work has been carried out. For instance, a comparison of the stability of a multi-enzyme reaction such as ethanol production by yeast cells immobilised by the different methods available for immobilising cells would be interesting. Five main methods have been used for cell immobilisation: entrapment, microencapsulation, covalent coupling, adsorption, and aggregation.

6.4.1 Entrapment

Entrapment (occlusion) inside biochemically inert hydrogels is the most common method of cell immobilisation, with polyacrylamide being the most popular support material. This is probably because entrapment is applicable to nearly all types of cell, whereas the other methods are often more specific or have grave disadvantages. Cells can be entrapped either by covalent bonding such as in the polymerisation of polyacrylamide; by ionic forces such as in calcium alginate immobilisation; or by precipitation by pH, temperature or solvent changes. The cells may be enclosed in a preformed three-dimensional matrix, or by storing the monomer and cells together and polymerising (O'Driscoll, 1976). The pores in the matrix must be sufficiently large to allow reasonably free diffusion of substrate into and product out of the support, but small enough to completely retain the entrapped cells.

6.4.1.1 ENTRAPMENT IN POLYACRYLAMIDE

Some of the first examples of cell immobilisation were by entrapment in polyacrylamide gel. In 1966 Mosbach and Mosbach immobilised the powdered thali of the lichen *Umbilicara pustulata* whose orsellinic acid decarboxylase activity was retained during periodic tests over 3 months. Updike *et al* (1969) next entrapped *Escherichia coli* and *Tetrahymena pyriformis,* and the protozoa could be observed swimming inside the gel for several days. In two papers (Franks (1971 1972) described the catabolism of arginine using immobilised *Streptococcus faecalis,* this being a multi-enzyme continuous process which was stable for 11 days. In addition, conversion of glucose into glutamic acid by a muli-enzyme pathway of immobilised *Corynebacterium glutamicum* was reported by Slowinksi and Charm (1973). Only in 1974 was the first industrial application of immobilised cells reported; the production of aspartic acid by the Tanabe Seiyaku Company in

Japan (Tosa *et al,* 1974; Chibata *et al,* 1974a; Chibata *et al,* 1976). Subsequently, the same company has developed a similar process to make L-malic acid (Yamamato *et al,* 1976; Yamamato *et al,* 1977), and the industrial production of prednisolone in Sweden using cells entrapped in polyacrylamide gel began in 1978 (Mosbach, K. personal communication).

A variety of other potentially useful laboratory scale cell immobilisation projects have also been described especially by Chibata's group. These include the immobilisation of *Achromobacter liquidum* so as to form urocanic acid (Yamamoto *et al,* 1974a). In this process cells were treated at 70°C for 30 min. to inactivate urocanase, and the detergent cetyl trimethyl ammonium bromide was added to facilitate substrate and product transfer across the cell wall.. Divalent ions had to be added to the substrate to replace that lost by gradual leakage from the cells due to the loss of membrane integrity. Similarly, *Pseudomonas putida* and *Escherichia coli* have been entrapped in polyacrylamide and used to form L-citrilline and tryptophane respectively (Yamamoto *et al,* 1974b; Chibata *et al,* 1974b), the former cells losing no arginine deimidase activity during one month's use. Immobilised *Escherichia coli* cells have been used to form 5-hydroxy-L-tryptothan from 5-hydroxyindole and serine or ammonium pyruvate (Chibata *et al,* 1974c). *Streptomyces griseus* and *Achromobacter* cells possessing glucose isomerase and NADP kinase activities respectively have also been immobilised in polyacrylamide; both processes requiring the addition of divalent ions to the substrate, Mg^{2+} and Co^{2+} to enhance and stabilise the glucose isomerase, and Mn^{2+} or Mg^{2+} to activate NADP formation (Chibata *et al,* 1974d; Chibata *et al,* 1975a). Lastly *Pseudomonas tagnei* or *Pseudomonas dacunhae* immobilised in polyacrylamide have been used to convert aspartic acid into *L-alanine* in good yield (Chibata *et al,* 1975b), and immobilised *Escherischia coli* cells used to convert penicillin into 6-APA with a 78% yield and a half-life of 42 days at 30°C or 17 days at 40°C. (Sato *et al,* 1976).

Among other workers Martin and Perlman (1976a and b) co-immobilised *Gluconobacter meganogenus* and a *Pseudomonas syringae* to carry out the two-step conversion of sorbose to 2-keto-L-gulånic acid, an important intermediate in Vitamin C manufacture. The L-sorbose dehydrogenase of *G. meganogenus* first forms L-sorbosone and then the sorbosone oxidase activity of *P. syringae* oxidises the sorvosone to ketogulonic acid. Saif *et al* (1975) immobilised *Escherchia freundii* whose acid phosphatase activity had a half-life of 120 days, Kinoshita *et al* (1975) used immobilised *Achromobacter guttatus* to hydrolyse ϵ-aminocaproic acid cyclic dimers, and Ohmiya *et al*

(1977) used *Lactobacillus bulgaricus, Escherichia coli,* and *Kluyveromyces lactis* to hydrolyse lactose in milk, the kinetic behaviour of the cells being virtually unchanged upon immobilisation. Recently Pache (1978) has used the β-lactamase activity of immobilised *Escherichia coli* cells to study the effect of β-lactamase inhibitors. Other reports include the formation of L-tyrosine and L-dopa using *Erwinia herbicola* (Yamada *et al,* 1975; Kumagaya *et al,* 1976); the oxidation of unsubstituted alditols and aldose diethyl dithioacetals (Schnarr *et al,* 1977), and tryptophan synthesis by immobilised *Escherichia coli,* 56% of the original activity being retained after immobilisation giving a productivity of 0.2g/L/h and a half-life of 30 days (Bang *et al,* 1978) *Mycelia* of *Mortierella vinacea* have been entrapped in polyacrylamide and used in a fluidised bed reactor to hydrolyse the oligosaccharides of soybean milk which are thought to be partially responsible for causing flatulence (Thananunkel *et al,* 1976).

Immobilised cells have also been used to affect steroid conversions. Skyrabin *et al* (1974) have described $\Delta^{1,2}$ dehydrogenation, C-20 ketone reduction, deacetylations, and IIα and IIβ hydroxylations using *Mycobacterium globiforme* and *Tieghemella orchedis.* Polyacrylamide entrapped *Mycobacterium globiforme* or *Corynebacterium simplex* transformed Δ^5 - 3β - hydroxy, Δ^5- 3β acetoxy and Δ^4 - 3 ketosteroids to $\Delta^{1,4}$ - 3 - ketosteroids in 70–80% yield (Voishvillo *etal,* 1976). Venkatasubramanian *et al* (1975) described conversion of cortisol to prednisolone using *Corynebacterium simplex* and *Arthobacter simplex,* 40% of this activity being lost on immobilisation. The half-life was only 5.5 days but could be lengthened by regeneration of the cells after addition of nutrients. The Δ' dehydrogenation of Reichstein compound S using polyacrylamide immobilised *Pseudomonas testosteroni* in a stirred reactor was performed by Yang and Studebaker (1978). The operational half-life was 103 h and phenazine methosulphate was used as an electron acceptor to increase the rate of reaction. Finally there is the industrial process for producing prednisolone developed by Mosbach and co-workers (Larsson *et al,* 1976; Ohlson *et al,* 1978; Ohlson *et al,* 1979).

Variations on the basic polyacrylamide immobilisation method have also been developed; for instance the Kyowa Yuka Co. have patented the co-polymerisation of acrylamide, acrylic acid, the metal salt of acrylic acid, cross-linking agents and polymerisation initiator in the presence of the cell. The resulting highly active gel has good mechanical strength (Japanese patent No. 138414).

In most cases the slurry of cells in acrylamide is polymerised into a block, and then mechanically granulated often by forcing the gel through a wire mesh to form discrete pellets of irregular shapes and sizes. In columns these particles pack irregularly and cause uneven flow and the development of relatively high flow induced pressure drops so, necessitating the use of several columns in series or sectionalised columns. (Sato *et al,* 1975). A continuous method for formation of regular sized pellets has recently been described (Kostner and Mandel, 1976) and these pellets should generate much smaller pressure drops. Mass-transfer rates inside porous gellified media are influenced by both the partition coefficient of the substrate between the bulk substrate solution and gel phase, and by the effective diffusion coefficient of the substrate inside the gel. Internal diffusion coefficients and therefore substrate transfer rates increase with an increase in the pore size, but decrease with increases in the molecular weight, the degree of hydration of the solute molecules, and the concentration of immobilised cells. (Yasuda *et al,* 1969; Yasuda *et al,* 1971; Nakanishi *et al,* 1977, and Mavituna and Sinclair, 1978). The pore size of the gel decreases but the mechanical strength of the gel increases in proportion to the (acrylamide concentration)$^{1/2}$ (B.D.H. pamphlet *Cyanogum*) and on the degree of cross-linking of the gel.

Polyacrylamide immobilised cell particles are easily abraded in stirred reactors. Polymerisation is also hindered by high concentrations of entrapped cells, and some loss of cell integrity occurs during immobilisation. Also much of the activity of the immobilised cells is often lost during polymerisation because of the denaturing effect of the acrylamide monomer and the free radicals and heat generated during polymerisation. This problem is analogous to that encountered in the development of biocompatible inplanted polymers for use as tissue substitutes in surgery. Here it is often not the polymer which is cytotoxic but the presence of necessary additives such as catalysts and anti-oxidants (Bagnall, 1978).

6.4.1.2 ENTRAPMENT BY OTHER METHODS

Entrapment of cells in agar gel is an obvious method of cell immobilisation, but has been little used presumably because of the poor mechanical strength of the gel and the need to heat the cell slurry before gelation will take place upon subsequent cooling. Examples are the conversion of malate to hydrogen using *Rhodospirillum rubrum* in an aerobic illuminated flowthrough reactor (US Patent 672,631), the entrapment of *Lactobacillus bulgaricus* reported by Miyata *et al* (1975), the production of pantothenic acid by *Escherichia coli* (Kawabata and Demain, 1978), and the entrapment of cells with glucose isomerase activity in gelatine agar and konjac gels (Kanno *et al,* 1978).

Immobilisation of cells, in calcium alginate gels is an example of coarctation, the polymerisation of polyelectrolytes by multivalent ions. Alginate, which is extracted from seaweeds, can be gelled by calcium ions, forming a linear block co-polymer of β-D-mannuronic and α-L-guluronic acids. A uniform, spherical, highly micro-porous structure is easily formed which retains molecules and particles larger than the pores. The pores are large and continuous, such that substrate molecules may diffuse freely throughout the pellet. The mechanical properties of calcium alginate gels are related to the distribution of the mannuronic and guluronic residues and to the molecular weight and degree of dispersity within the sample; for instance, the highest breaking strengths were obtained using alginates having a high concentration of polyguluronic blocks (Haug *et al,* 1974). The alginate immobilisation technique is versatile allowing immobilisation of many different types of cells and organelles, and it enables stable immobilised cell preparations of high activity and effectiveness factor to be formed cheaply and simply with little loss of activity (Kierstan and Bucke, 1977; Cheetham *et al,* 1979). Other workers have used alginate immobilised cells; for instance, Dallyn *et al* 1977) studied the inactivation of alginate immobilised spores in viscous solutions at ultra-high temperatures, and Hackel *et al* (1975) found that immobilised *Candida tropicalis* hydrolysed phenol to CO_2 and H_2O at 28% of the rate of the free cells, while recently White and Portno (1978) rendered *Saccharomyces cerevisiae* sufficiently non-flocculant to use in continuous tower fermenters. No microbial contamination and no deterioration in fermentation ability occurred during 7 months' use, but the freshly formed cells were not trapped by the gel. The non-denaturing properties of the alginate, and the resistance of the pellets to CO_2 generation inside the entrapped cells, make this method especially useful for multi-enzyme conversions such as ethanol formation (Fig. 6.3). Large-scale use appears to be possible because pellets can be mass produced, and when used in packed-bed columns only low intrinsic pressure drops could be measured, chiefly because of the even size and spherical shape of the alginate pellets (Cheetham *et al,* 1979; Cheetham, 1979). Such excellent resistance to hydrostatic pressures is not unexpected, because the biological role of alginate is to provide mechanical support for seaweed tissues. No channelling, abrasion of pellets or compression were observed during use in columns, but abrasion was noticed when pellets were use in stirred reactors, the abrasion increasing with the pellet concentration and stirrer speed. A further disadvantage of the use of alginate as an immobilisation support is that moderate concentrations of calcium chelating agents such as phosphates, EDTA, and certain cations such as Mg^{2+} or K^+

disrupt the gel by solubilising the bound Ca^{2+}. Leakage of the entrapped cells from the gel pellets also occurred at a low rate, especially when cell division within the pellets takes place and when the pellets are used in stirred or shaken vessels.

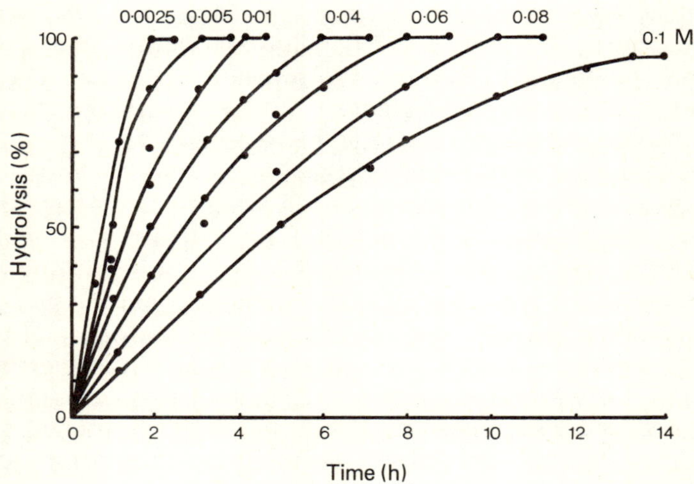

Fig. 6.3 – Hydrolysis of acetyl-L-methioine using aminoacylase pellets. Fifty ml samples of acetyl-L-methionine substrate was recirculated through the column (V_s = 27 h^{-1}) (Hirano *et al,* 1977). The half-life of the cells was 250 h (Kierstan and Buche, 1977).

Recently a modified form of alginate immobilisation has been described (Klein *et al,* 1978). A co-polymer of alginate and epoxide containing the entrapped cells was formed and these pellets were reduced in size by drying; then the alginate was dissolved away by phosphate buffer to leave an open pored, rigid immobilised cell pellet. Thus a more complex immobilisation technique has been used which involves drying and consequent loss of the immobilised cell activities. Also cell division within the pellets was not observed, and so *in situ* cell division may only occur inside flexible support materials.

Reconstituted collagen has been widely used as an enzyme and cell immobilisation support (Vieth *et al,* 1973). Cells are mixed with collagen and a membrane cast, dried, and tanned with glutaraldehyde. Alternatively, glutaraldehyde may be added directly to the cell-collagen dispersion prior to casting (Vieth and Venkatasubramanian, 1976). Immobilisation has been described as being due to a network of non-covalent interactions (Venkatasubramanian *et al,* 1974), but reacently ϵ-amino lysine residues located within the helical portion of the tropocollagen molecule have been implicated as being the primary receptors and so cell immobilisation using collagen may therefore be better described as an adsorption technique. Using this method, the glucose isomerase activity of *Streptomyces phaeochromogenes* has been utilized in a continuous flow column or in spirally wound biocatalytic reactor modules (Vieth and Venkatasubramanian, 1976). A number of other activities have also been immobilised such as hydrocortisone and cholesterol conversion to prednisolone; citric, aspartic, gluconic and 2-ketogluconic acid formation; candicidien synthesis; biophotolysis of water, plutonium removal (Venkatasubramanian, 1977). Yamada *et al* (1977) have also used collagen as a support material immobilising *Erwinia herbicola* with 88% retention of the β-tyrosinase activity.

Cellulose and cellulose derivatives have also been widely used as cell immobilisation supports. For instance Shimizu *et al* (1975) entrapped *Brevibacterium ammonigenes* in cellophane tubing and silastic resin, and used a five enzyme pathway of the cells to form co-enzyme A. Leuscher entrapped *Aspergillus oryzae* in collodian so as to utilize the aminoacylase activity (German Patent 1,227,855), and Kolarik *et al* (1974) immobilised cells in cellulose acetate; however, the cells leaked out of the support and diffusional effects restricted the glucose isomerase activity of the cells. The penicillin acylase and cephalexin synthesising activities of *Escherichia coli* were also retained when the cells were entrapped in cellulose triacetate (Dinelli, 1972; Marconi *et al,* 1975). The most extensive study of cellulose as a cell immobilisation support has been by Linko and co-workers (1977), who suspended dried *Actinoplanes missouriensis* in cellulose dissolved in N-ethyl pyridinium chloride and dimethyl formamide and precipitated them as beads by contacting with water. The glucose isomerase retained 40–60% of its activity after immobilisation to give an immobilised activity of 50–350 units/g of cell fibre. Cell leakage was prevented by glutaraldehyde treatment. When used in a continuous column at pH 7.5 and 60°C, the cells had half-life of 45 days and achieved 40% conversion.

A variety of artificial polymers have also been used as entrapment supports; for instance Guttag (1973) has entrapped cells in hydrophilic homopolymers of the lower alkyl, hydroxyl, alkyl or hydroxyloxy alkyl, acetate or methacrylates, and Yoshino *et al* (1975) formed polymers such as polycrylonitrile, polymethacronitrile, and cellulose and at the same time injecting an enzyme or cell suspension (*Bacillus megalerium* and *Fasarium solani*) into the fibre. No activity was lost during polymirisation. Kimura *et al* (1978) immobilised *Hansenula jadinii* in a derivative of polyethylene glycol, hydroxyethyleneacrylate, the cells forming ATP which phosphorylates CMP, which is then used to produce CDP-choline. The glucose isomerase activity of *Streptomyces phaechromogenes* cells has been entrapped by radiation initiated polymerisation of 2-hydroxyethyl methacrylate (Kumakura *et al,* 1978). High concentrations of cells could be immobilised with no leakage of cells, and the activity of the cells could be further increased by coentrapment of particulate materials. A novel technique is reported by Petrie *et al* (1978), who mixed *Escherichia coli* with glutaraldehyde and bovine serum albumin, spread it as a film, dried it and ground it. The β-galactosidase activity of the cells had a reduced K_m after this treatment.

Recently, entrapment of cells in kappa-carageenan, by mixing cells with the carageenan at 45–60°C and then cooling, cross-linking with glutaraldehyde or hexamethylene diamine and granulating the block of gel has been developed (Tanaka *et al,* 1979, Nishida *et al,* 1979). Alternatively the gel can be further strengthened by incorporation of locust bean gum. This support is to replace polyacrylamide in the Tanube Co. industrial processes because of their increased productivity.

Finally the entrapment of heat treated *Streptomyces* cells containing glucose isomerase in polyester sacs has been described in detail by Ghose and Subhash (1978). External film and internal pore diffusional resistances to diffusion of the hydrolysed microcrystalline cellulose were negligible. The half-life of the immobilised enzyme activity was 18.6 days compared with one day for the free enzyme, the activity of the preparation being 250–300 μmoles/min/mg of dry cells.

6.4.2 Microencapsulation

Microencapsulated cells are retained inside a single semi-permeable membrane, such as collodian or silicone, which allows substrate and product to diffuse relatively freely. For instance, Mohan and Li (1975) mixed a buffered suspension of viable *Micrococcus denitrifricans* cells with oil, surfactant, a membrane strengthener and anion transport facilitator, to form an emulsion, each droplet being 20–40 μm in diameter and containing 500–600 cells. When substrate was added the reaction rate was still limited by the substrate transfer rate, but 78% of the initial activity was retained after 120h; also, after 5 days, phase contrast microscopy showed that no cell lysis had occurred. Microencapsulation has been extensively used by Chang and co-workers, for instance see Chang (1971); but the droplets are probably too fragile to be used industrially. Sedimentation of cells within the fluid core of the capsule may also occur. Therefore this technique appears to be limited to medical and analytical applications.

6.4.3 Covalent Coupling

Covalent coupling is the direct linkage of cells to an activated support. Linkage can be to any reactive component of the cell surface, for instance, the amino, carboxyl, sulphydryl, hydroxyl, imidazole or phenol groups of proteins can be utilised. This technique has the advantage that cells are linked to a uniform surface by a bond which is stable for long periods, so that cell leakage is minimised. Unlike covalent immobilisation of enzymes no chemical modification of the enzyme activities of the immobilised cell occurs, unless they are localised on the outside of the cell.

In 1974 Chipley described the immobilisation of a number of bacteria by carbodiimide bonds formed between the cells and agarose adipic hydrazide beads, but both the activity and viability of the cells were lost. Later, *Brevibacterium ammoniagenes* was immobilised in a co-polymer of ethylene with malic anhydride, but again activity was lost (Shimizu, 1975). An especially interesting cell immobilisation technique is the immobilisation of a variety of cell types to a polymeric metal hydroxide precipitate formed from aqueous solution of Ti^{4+} and Zr^{4+} chlorides (Kennedy *et al,* 1976). Some form of partial covalent coupling involving interactions between the cell and the metal hydroxides was presumed by the authors, and the cells still respired after immobilisation. Recently Jack and Zajic (1977) immobilised *Micrococcus lutens* to carboxylate functionalised agarose supports. Presumably because the cells were not exposed to free carbodiimide, the urocanic acid-forming activity was still retained after cell viability had been lost. Navarro and Durrand (1977) immobilised *Saccharomyces carlsbergensis* to porous glass beads activated with glutaraldehyde, but if the pores in the glass were large enough adsorption of the cells also occurred. In both preparations the conversion of glucose to ethanol increased and conversion to CO_2 decreased on immobilisation. A further method of covalently immobilising cells is by treating them with glutaraldehyde then with glycidyl methacrylate monomer followed by polymerisation (UK Patent 1,478,272).

6.4.4 Adsorption

Adsorption of cells involves the formation of a number of ionic and hydrogen bonds between the cell surface and the support material. Adsorption depends on the charged nature of the cell wall, which is chiefly determined by the distribution of carboxyl and amino groups. Adsorption is a mild, non-specific process, but stability is limited by the rate of desorption of cells from the support (Marcipar *et al,* 1978). This problem is especially severe when changes in pH or ionic strength occur, when multiplication of the adsorbed cells occurs (Hattori, 1972), or when cells are sheared from the surface of the support owing to rapid fluid flow, contact with gas bubbles or other particles. Thus precise control of the quality of the substrate or cross-linking of the cells in place after adsorption is required. A further disadvantage of adsorption methods is that usually only a low proportion of the total number of cells exposed to the support can be immobilised if immobilised preparations of high activity are required. As with covalently coupled cells internal diffusional restrictions can be minimised by immobilising the cells to the outside of non-porous supports; however, the activity per gram of immobilised cell will be low and cells are easily abraded off the support. An important practical advantage is the ability to renew the catalyst by desorption of inactive cells, followed by readjustment of the pH or ionic strength and adsorption of fresh cells.

One of the very first attempts to immobilise cells was performed by Hattori and Fursaka (1961), who adsorbed *Escherichia coli* and *Azobacter agile* to Dowex-1 and measured the oxidation of succinic acid. Subsequently, Johnson and Ceigler (1969) immobilised a variety of micro-organisms on several ion-exchange resins, best results being obtained with ECTEOLA-cellulose. Immobilised spores were found to be less active, but more stable than vegetative cells. Fujii *et al* (1973) adsorbed an *Achromobacter* sp. and *Bacillus megaterium* to DEAE cellulose, using them for ampicillin synthesis and acylation of the cephalosporin nucleus. Ishinatsu *et al* (1976) adsorbed *Actinomycetes* cells to anion exchange resins and used the glucose isomerase activities of the immobilised cells.

A novel use of adsorbed cells is in the culture of mammalian cells in monolayers on the surface of small positively charged beads of Sephadex suspended in the culture medium by gentle stirring; Advantages are the high cell yields, ease of handling and control, and the ability to carry out large-scale continuous culturing (van Wezel, 1967). This system is now marketed by Pharmacia under the trade name Cytodex 1. The adsorption of cells to ion-exchange columns is also being used to measure the numbers of micro-organisms in foodstuffs (British Food Manufacturing Industries Research Reports 245 and 276). *Pseudomonas* adsorbed to anthracite has been used to detoxify phenol in a tapered fluidised bed requiring a residence time of only a few minutes to reduce the phenol concentration to 10–50 ppb. Conversion rates were limited by oxygen supply and cell division of the attached cells occurred (Scott and Hancher, 1976).

6.4.5 Aggregated Cells

Rather than immobilise cells in a thin layer on the surface of the support material, cells can be pelleted, cross-linked or flocculated by cationic or anionic polyelectrolytes into a cell aggregate. Such cell clumps can be regarded as carrierless immobilised cells, and have similar practical advantages to the other forms of immobilised cells described earlier. Examples are found in the production of high fructose syrups, using cells possessing high glucose isomerase activities because the cost of this product is now so low that a very cheap immobilised biocatalyst must be used. Instances are natural mould pellets (Kobayashi and Suzuki, 1976), pelleted cells (Regan, 1977), cells cross-linked with glutaraldehyde (Poulson and Zittan, 1976; Miles Labs. Inc., 1974), or with diazotised diamino compounds (U.K. Patent 1,376,983); heat treated *Streptomyces albus* cells (Takasaki, 1969) and flocculated cells (Long, 1977). Novo Industries use ruptured *Bacillus coagulans* reacted with glutaraldehyde or aggregated with flocculant, and then cross-linked with glutaraldehyde before granulation of the aggregate. Release of the intracellular contents is essential for good cross-linking of the cell debris (U.K. Patent 1,516,704). The Mitsu Seito Co. have patented the use of immobilised *Actinomyces* glucose isomerase in which the gels are mixed with gelatin or casein and then reacted with glutaraldehyde in acetone (Japanese Patent 144639).

A non-glucose isomerase use of aggregated cells is the hydrolysis of raffinose from beet sugar with the α–galactosidase of *Mortierella vinacea* mycelia, (Meguro and Konish, 1977). Industrial processes using pellet bound α–galatosidase have been used since 1968 by the Hokkaido Sugar Co. Ltd. in Japan and since 1974 by the Great Western Sugar Co. in the U.S.A. (Obara and Hashimoto, 1976/7).

6.4.6 Miscellaneous Methods

Biospecific adsorption has great promise. It offers the advantages of very specific, tight, non-covalent binding under non-denaturing conditions, to a variety of specific materials. Natural materials such as antibodies or lectins are possible reagents, but both are expensive and available only in small quantities. An example is the lectin concanavalin A bound to magnetite. The lectin selectively interacts with the α–mannans in the cell walls of *Saccharomyces cereviase* and *Candida utilis,* so enabling magnetic removal of the cells (Horrisberger, 1976). Very recently Pharmindustrie of France have marketed a commercial absorbent lectin–Ultrogel, which is lectin immobilised to Ultorgel AcA22 by glutaraldehyde.

A novel technique is the immobilisation of enzyme onto the surface of cells. By this method glucoamylase has been coupled to $TiCl_4$ treated *Saccharomyces cerevisiae* (Hough and Lyons, 1972);

also aminoacylase and glucose oxidase have been immobilised on natural mycelial pellets of *Aspergillus* sps. using glutaraldehyde and albumin (Hirano *et al,* 1977 b; Karube *et al,* 1977d). In the latter case 95% of the original activity was retained, no leakage of enzyme occurred, and complete oxidation of glucose in a recycle reactor was achieved. Hirano *et al,* (1975) had earlier used a similar technique of cross-linking cells with glutaraldehyde and albumin to immobilise the aminoacylase activity of *Aspergillus ochraceus.*

6.5 THE KINETICS OF IMMOBILISED CELLS

The quantification and prediction of the performance of complex systems such as immobilised cells can be achieved by using mathematical models. Modelling may enable the prediction of the behaviour of large-scale reactors, give an understanding of the way in which cellular processes act as an integrated whole, identify specific areas in which improvements can be made, and facilitate the transition from laboratory to industrial practice. Equations should be simple and general, that is apply to a number of micro-organism-substrate systems even though the cells in different systems will not be genetically, morphologically and biochemically identical.

A further problem is that academic enzymology usually concentrates upon initial rate measurements of the formation of product and consumption of substrate made in dilute aqueous solutions, whereas in industrial processes complete conversion of concentrated substrate solutions is often required and the same enzyme is often used continuously over periods of weeks or months. Therefore the time taken to achieve 95% conversion or the effectiveness factor achieved under defined conditions are often more relevant experimental parameters. Another difficulty arises because enzymes do not affect the position of equilibrium in a chemical reaction only when present at very low concentrations, whereas in many commercially important reactions the concentrations of enzymes used are of the same order as the substrate concentrations and so much of the substrate is present as an enzyme-substrate complex. Furthermore it should not be disregarded that enzymes may fulfil other functions in addition to their catalytic roles, for instance to store metabolites and to protect these metabolites.

The intrinsic kinetics of immobilised cells may be defined as the behaviour of the enzyme activities of the free cell. The inherent and effective kinetics may be similarly defined as the behaviour of the immobilised cell in the absence of diffusional limitations, and when diffusional limitations are significant respectively. The kinetics of immobilised cells are more complex than the expressions describing single enzyme activities, but are simpler than the kinetics of fermentations because usually no growth occurs.

In heterogenous catalytic systems such as immobilised cells, the overall rate of reaction is determined not only by factors such as pH, temperature and substrate concentration, but by the rates of ion, heat and substrate transfer. Six main features have been implicated in the modification of immobilised enzyme kinetics (Goldstein, 1976), and many of these effects will also apply to immobilised cells. These six effects are (a) conformational effects caused by modification of the enzyme during immobilisation, (b) steric effects by orientation of the enzyme after immobilisation, (c) partitioning of substrate, product or ions between the bulk solution and the immobilisation support owing to hydrophobic or electrostatic interactions, (d) micro-environmental effects, which reflect the fact that enzyme-substrate interactions occur in a changed micro-environment because when a cell is immobilised its immediate environment is greatly influenced by the physical properties of the support material, (e) external diffusional restrictions on substrate and product transfer in the unstirred boundary (Nernst) layer around each particle of immobilised cells, and (f) internal diffusional restrictions on substrate or product transfer within the support material. As the enzymes of immobilised cells do not directly interact with the support material micro-environment, conformational and steric effects are unlikely. Partitioning effects may occur because cells usually have an overall negative charge which may cause shifts in the pH and substrate concentration optima of the enzyme activities of the immobilised cells. For instance the optimum pH of *Escherichia coli* and *Azobacter agilis* cells immobilised to anion exchange resins were higher than for the same cells in suspension (Hattori, 1973). Such shifts may be disdavantageous, but can often be usefully exploited to allow operation in previously unpractical conditions.

Diffusional effects are the most important modifiers of the inherent immobilised cell kinetics especially in the case of fast enzymic reactions or where low substrate concentrations, highly porous, very active, or large immobilised cell particles are used. Then simultaneous diffusion and reaction occur such as described for the whole cell invertase of *Saccharomyces pastorianus* by Toda (1975) and Toda and Shoda (1975) who immobilised cells in spherical agar pellets by injecting into toluene or trichloroethylene and found the internal mass-transfer coefficient to be correlated with particle size, intra-particle enzyme concentration, and external substrate concentration. The kinetics of the less stable lactase activity of *Escherichia coli* were also investigated (Toda, 1975).

External diffusional restrictions can be minimised by increasing the rate of agitation in a stirred reactor or by increasing the flow rate through a tubular reactor. Internal diffusional restrictions can be detected by a lowering in the apparent activation energy for the reaction, and may be expected to be a major influence on immobilised cell activities because immobilised cell particles are usually much larger than the corresponding immobilised enzyme preparations. However, because cells are much larger than enzymes, support materials with very much larger pore sizes may be used for all immobilisation, which allows much faster diffusion of substrates within the cell supports. These large pores also enable the use of much larger substrate molecules. Internal diffusional restrictions are represented by the Thiele modulus, values less than 1.0 being obtained in diffusionally rather than reaction controlled systems. In general a balance has to be achieved between immobilised cell preparations of a high activity per unit reactor volume and with a consequently low effectiveness factor, and preparations of low activity and high effectiveness factor. Of course where a high activity is achieved, the effectiveness factor will be high near the surface of the support material and low in the core of the particle because of internal diffusional effects. Internal diffusional restrictions could be reduced by using irregular support particles with a relatively high ratio of surface area to volume, but of course these would rapidly abrade in a stirred reactor and cause uneven flow and compression in packed-bed reactors.

The kinetics of immobilised cell preparations will usually be more complex than those of the corresponding immobilised enzymes because of two additional effects. These are the possibility of a limited amount of cell division by the immobilised cells, and the presence of an additional diffusion barrier created by the presence of an osmotically intact cell wall and cytoplasmic membrane, the extent of which will vary greatly with the structure of the cell wall and membrane and the type of substrate used. Moreover, coupled enzymes and even complete metabolic pathways of the immobilised cells can be utilised, together with the regeneration of co-factors and use of active transport mechanisms. The presence of a further diffusion barrier at the cell surface is clearly indicated by the presence of an intact osmotic barrier, and by the ability of the cell to respond to a mild osmotic shock. The effect of this diffusion barrier can be quantitated by the increases in immobilised cell activities after cell lysis, especially when surface active agents have been used to overcome mass-transfer limitations. For instance, in the formation of urocanic acid using *Achromobacter liquidum* (Yamamoto *et al,* 1974b), co-enzyme A using *Brevibacterium am-*

moniagenes (Shimizu *et al,* 1975) and malic acid also using *B. ammoniagenes* (Yamamoto *et al,* 1976), detergents are no longer required to facilitate substrate transfer, and divalent ions have to be added to the substrate, presumably to replace those lost by leakage from the cell.

Thus the effective kinetic parameters of the enzyme activities of immobilised cells such as V'_{max} or K'_m are seldom closely related to the intrinsic or inherent kinetics of these enzymes. Consequently, the Michaelis-Menten equation is modified to:–

$$\nu = \eta \frac{(V_{max}\, S)}{(K_m + S)} \qquad (6.1)$$

where ν is the observed rate of reaction, S the bulk substrate concentration, V_{max} the maximum rate of reaction, K_m the Michaelis-Menten constant and η the effectiveness factor, which is defined as the ratio of the observed reaction rate to that achieved in the absence of diffusional limitations. For instance, Hirano *et al* (1977a) have used an integrated form of the Michaelis-Menten equation to describe the kinetics of the aminoacylase activity of pellets of *Aspergillus ochraceus* used in columns with substrate recycled through the column. The K_m was decreased by increasing the space velocity, and complete hydrolysis of substrate was achieved by using substrate concentrations less than 0.08M (Fig. 6.4). In general, effectiveness factors in excess of 1.0 can be obtained when the permeability of the

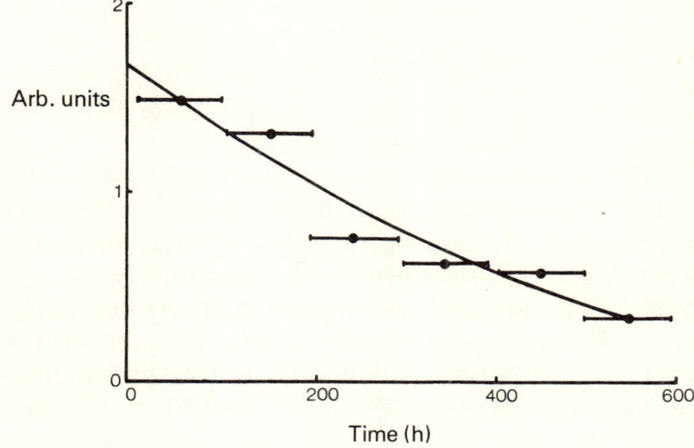

Fig. 6.4 – The rate of ethanol production by immobilised *S. cerevisiae*. The half-life of the cells was 250h (Kierstan and Buche, 1977).

walls of the immobilised cells is increased and when the immobilised cell concentration is increased by *in situ* cell division even though the new cells are usually not evenly dispersed inside the immobilisation support.

Immobilised cell support particles are usually many orders of magnitude larger than the immobilised cells and so because of the great number of immobilised cells present individual differences can be ignored and average properties used. The kinetics of immobilised cell activities may approximate to the cores of encapsulated enzymes (Atkinson, 1974), or more closely to microbial flocs (Atkinson and Daod, 1976), as a cell floc consists of cells uniformly surrounded by an intercellular gel, this gel occupying 20–40% of the floc volume.

A model for a single metabolising cell consists of the cell surrounded by outer concentric transport zones. If substrate transfer occurs entirely by diffusion, then assuming unidirectional flow and an absence of partitioning effects the substrate flux due to diffusion is described by Fick's law. As the cell can be regarded as a substrate sink activity can be reduced to a form of Michaelis-Menten kinetics where the rate coefficients for the individual steps are condensed into the coefficients α_R and β_R. This is because the kinetics of the complete metabolic pathway are chiefly determined by the rate limiting reactions. Thus when immobilised cells are used continuously an equilibrium is set up: where:

$$a_m \left(-D_m \left| \frac{\mathrm{d}s}{\mathrm{d}x} \quad a_m \right. \right)^{\dagger} = \left[\frac{V_R \ \alpha_R \ S''}{a_R \ (\beta_R + S'')} \right] a_R \qquad (6.2)$$

($\dagger a_m$ indicates that the concentration gradient is evaluated at the external surface of the micro-organism).

where S'' is the substrate concentration in the metabolic zone, D_m is the diffusion coefficient of the substrate, ds/dx represents the substrate concentration gradient a_m and V_m are the external surface area and volume of single micro-organisms respectively.

This model holds for smaller molecules such as oxygen. For larger substrates, particularly those which act as carbon sources and are transported by permeases, the micro-organism is surrounded by two concentric transport zones, transport through the outer zone being diffusion mediated and transport through the inner zone being mediated by the premease which is powered by cellular energy reserves. Permease mediated transport is not affected by the bulk substrate concentration but is described by a Michaelis-Menten or

Langmuir isotherm type relationship (Cohen and Monod, 1957). A steady state is achieved in which the rate of consumption of substrate by the cell is equal to, and limited by, the flux of substrate through each successive transport zone. Thus:

$$a_m \left(-D_m \frac{ds}{dx} \Big| \, a_m \right) = a_p \left(\frac{\alpha_p \, S'}{\beta_p + S'} \right) = V_R \left(\frac{\alpha_R \, S''}{\beta_R + S''} \right) \quad (6.3)$$

where S' is the substrate concentration external to the permease region, α_p and β_p are coefficients and a is the external area of the permease zone (Atkinson, 1974).

Provided that the cell mass functions and biological properties remain constant with time a differential mass balance for the immobilised cell preparation can be established, in which the rate of transport of substrate is equated to its removal by the cells and the appearance of product. That is:

$$D_e \left(\frac{d^2 S}{d x^2} \right) - \frac{a \, \alpha \, S}{\beta + S} = 0 \quad (6.4)$$

where a is the external surface area of viable micro-organisms per unit volume of immobilised cell pellet; α represents either α_p or $\frac{\alpha_m \; V_m}{\alpha_m}$, and β is a rate equation coefficient. That is when $x = L$ (the thickness of the immobilised cell preparation) $dc/dx = 0$, and when $x = 0, S = S'$ (Atkinson, 1974), D_e represents the effective molecular diffusion coefficient within the tortuous intrapellet pores and reflects the hydrogen bond mediated interactions of substrate with the support material and steric hindrance between substrate and the support (Nakanishi *et al*, 1977).

$D_e = \epsilon/\tau$ where ϵ is the porosity of the support, τ the tortuosity of the pores, and D is the normal bulk diffusion coefficient of the substrate. However, division of the entrapped cells causes complications. Often the mechanical strength of the immobilised cell preparation is weakened and in extreme cases the granules may be broken. Although the concentration of immobilised cells has been increased, the new cells are not evenly dispersed within the support material, which leads to increased diffusional restrictions. Furthermore, the presence of these additional cells reduces the effective diffusion coefficient of the substrate inside the support in proportion to the percentage of the gel cross-section occupied by cells (Mavituna and Sinclair, 1978).

The problems of getting product out of the immobilised cell preparation are very similar to those concerned with substrate movement into the cell. However, in many cases leakage of intermediary metabolites out of the cell may be greatly enhanced where the cell has lost its chemi-osmotic integrity.

As has been shown for immobilised enzymes where diffusional limitations exist in heterogeneous systems, inhibition of the enzyme activity of the immobilised cells affects both the inherent activity of the enzyme and the rate of depletion of substrate. Thus:

$$\nu' = \eta' \frac{\nu_{max} S}{K_m + S} \tag{6.5}$$

where η' is the effectiveness factor in the presence of inhibitor and ν' the initial rate of reactions in the presence of inhibitor. Thus when diffusional and chemical inhibition are operating at the same time, their combined effect is less than the sum of their separate effects, and so the effect of inhibitors may be masked under severe diffusional limitations, only becoming apparent when the diffusional restrictions are lifted. A common effect is the presence of very low concentrations of inhibitors or denaturants in the substrate which can cause a slow but progressive loss in the activity of the immobilised cells. When a porous support material is being used the activity is first lost on the outside of the pellet and then the cells located further inside the pellet are gradually inactivated.

Enzyme denaturation can be considered as a special case of enzyme inhibition, diffusional effects increasing the apparent activities of immobilised cell activities by a similar mechanism to the effect first described in immobilised enzymes by Ollis (1972). An example is the glucose isomerase activity of pelleted cells, where loss of activity was linear when diffusion limited activity, but at the end of operation when little activity remained, exponential decay was observed (Regan, 1977). For multi-enzyme systems, continued activity is also dependent on the preservation of the integrity of the cell membrane as measured by oxygen uptake, trypan blue exclusion and the increases in both enzyme activity and cell numbers when the immobilised cells are incubated with growth medium. Even when a complex series of enzymes are used in concert such as in the formation of ethanol from glucose, activity still declines exponentially, just like a single immobilised enzyme, presumaby because a single enzyme limits the overall stability of the metabolic pathway (Fig. 6.3). Other factors affecting the stability of immobilised cells are the possibility of microbial contamination, leakage of cells or enzymes and the effect of endogenous proteases. The operational stability of an immobilised cell preparation is usually expressed as a half-life, but more properly by a rate equation containing a deactivation constant. Of course if intermittent use is envisaged the storage stability is also important.

6.6 IMMOBILISED CELL REACTORS

Enzyme reactors differ from chemical reactors, chiefly because enzymes operate at one atmosphere pressure and comparatively little heat is generated or consumed during the reaction. As with chemical reactors, biochemical reactors are judged by their productivity and specificity.

If successful and practical use of immobilised cells is to be made, then an appropriate choice of reactor configuration must be made. This choice is intimately dependent on a variety of factors such as the chemical and physical properties of the substrate and product, the nature of the support material, the type of cell used, the method of immobilisation, and the type of enzyme reaction being performed.

Reactor configuration are classified according to the method of catalyst retention, the flow characteristics, and whether a continuous or discontinuous mode of operation is used. The concentration profiles of reactants within reactors of different configurations may vary appreciably; for example in a stirred reactor almost complete back-mixing is obtained with the substrate and product concentrations changing continuously with time, while in a packed-bed reactor in which plug flow is approximately achieved, the concentrations of substrate and product do not change quickly, but change continuously along the length of the column. These concentration profiles may affect both the activity and stability of the immobilised cells; for instance, endogenously produced inhibitors are constantly swept out of continuous reactors so that critically high concentrations are not usually reached. Batch, semi-batch, or continuous operation in stirred, thin-membrane, multi-chambered, stirred, packed-bed, expanded bed, or fluidised bed reactors are possible, with constant flow packed -bed and batch stirred reactors being the most popular. Thin membrane reactors include hollow fibre and ultrafiltration reactors. Continuous operation of the reactor has economic advantages, especially where a large output is required as capital and labour costs are decreased, and constant operation is most easily controlled and automated. However, catalyst replacement and regeneration are more difficult. The activity per unit reactor volume is very important, as this parameter chiefly determines the size of reactor required to achieve the desired productivity; it also influences operational factors such as the amount of bed compression encountered. When activity is calculated on the basis of reactor volume packed-bed reactors have an advantage over stirred reactors, as in the latter the particulate catalyst may only occupy 10% of the total reactor volume. The enzyme activity of the original cells, the weight of cells immobilised per litre of support material, the retention of activity after immobilisation and the effect of diffusional effects on the operational activity of the immobilised cells, the stabilities of the enzyme activities of the immobilised cells, and desired yield and purity of product, safety, and other operational requirements must be taken into account. For example a fluidised bed or stirred reactor is especially advantageous when good gas mixing, or efficient pH or temperature control, or use of high flow rates is required, or when substrate containing undissolved solids are to be used, but very friable support materials cannot be used and untapered fluidised beds can only be used over a narrow range of flow rates.

The conditions of immobilisation should be chosen so as to optimise reactor productivity which increases with both the activity and stability of the immobilised cells. This is a difficult area of research, because the greater the stability then the more apparent are the effects of very mild or unexpected stresses. Very often changes in one experimental parameter will have opposite effects on the activity and stability of the immobilised cell, necessitating a compromise based on the economics of the process. As an example, moderate changes in temperature will increase activity, but stability will decrease logarithmically. Such effects may be used to advantage, as in the synthesis of malic acid, where constant productivity can be maintained by gradually increasing the operational temperature, so as to compensate for the continuous decrease of fumarase activity with time (Yamamoto *et al,* 1976). Alternatively, the flow rate could have been gradually reduced because the percentage conversion is inversely proportional to flow rate. This regime would have the added advantage of gradually decreasing the pressure drop in a column which has pumped downwards but would also tend to increase the external diffusional limitations on reaction rates.

Table 6.1 shows the activities and stabilities of some of the more important types of immobilised cell, including the three industrial processes. This figure therefore indicates the approximate standards required for commercial success. Note also that the heat stabilities of immobilised cells always appear to be greater than the corresponding cells in suspension (Durrand and Navarro, 1978). The stabilities of the immobilised cells are much greater than the free cells (Fig. 6.1); but are not significantly different from the corresponding immobilised enzyme preparations, this point being best illustrated by the published glucose isomerase data.

Many other specialised factors also apply, for instance, loss of immobilised cell activity is often due to cell lysis, followed by leakage of intracellular contents, and in particular divalent ions. Inclusion of Mg^{2+} in the substrate is often sufficient to compensate, although a very limited amount of enzyme activity can be maintained by population effects, a few cells utilising the metabolites liberated from lysed cells. Using a porous compressible support in a packed-bed reactor, diffusional restrictions on reaction rate (which are reflected in lower effectiveness factors), and pressure drops in the immobilised cell column will both be increased by decreases in the size of the immobilised cell particles and the viscosity of the substrate. High effectiveness factors could be maintained throughout the length of the column by having the highest concentration of immobilised cells at the inlet end and gradually decreasing their concentration along the length of the column, with the minimum concentration being at the exit end of the column. This approach would be especially useful when the immobilised cell activity is stabilised by the presence of substrate, but would decrease the overall activity per unit volume of column volume.

Ideally, when scaling up a process, the model system should be geometrically and dynamically similar to the larger version. Use of packed columns allows continuous operation so decreasing labour costs (Fig. 6.1). Use of several columns in parallel is most satisfactory, especially when the individual columns are 'out of phase' with each other in a carousel arrangement, as this regime enables constant product quality and productivity to be maintained, despite the constantly declining productivity obtained from any individual column. Fewer problems with compression of the columns are encountered when using a number of smaller columns, but monitoring and control systems need to be more complex.

During use of immobilised cells multiple reactions often take place generating undesired side products. These may be caused chemical ly, by further metabolism of the desired product, by another enzyme or group of enzymes metabolising part of the substrate, or by a more complex mechanism. It may be that the by-product(s) are useful and can be easily purified and sold, but usually the aim is to stop or minimise their formation by a judicious choice of reactor configuration and reaction conditions. The yield of the reaction can be expressed either as a relative or operational yield. Relative yields represent the percentage of substrate molecules reacted which have formed the desired product. This parameter falls sharply with increasing conversion. Alternatively the operational yield is the percentage of substrate molecules fed into the reactor which have reacted to form the desired product. The selectivity of the reaction is defined as the number of molecules of desired product formed divided by the number of molecules of substrate transformed into unwanted product. Reactions in which back-mixing occur, such as in stirred tank reactors, generally have reduced selectivity because products are mixed with reactants, and the reactant concentration available for reaction is reduced. Yield and selectivity will therefore be greatest in tubular reactors approaching ideal plug flow.

6.7 CURRENT INDUSTRIAL IMMOBILISED CELL PROCESSES

Interest in immobilised cells has been stimulated by three processes which have been successfully used industrially, although none of them is operated on a large scale by the standards of the heavy chemical industry. The production of aspartic acid from fumaric acid using *Escherichia coli* cells immobilised in polyacrylamide gel is the best described immobilised cell process. The method was developed by Chibata and co-workers at the Tanabe Seiyaku Co. in Japan, and has been in production since 1973, producing 1,700 kg of aspartic acid per day per $1m^3$ column. Aspartic acid which is used in medicines and as a food acidulant in competition with citric acid had previously been produced industrially by fermentation or using soluble aspartase. The difficulties associated with such batch production methods were overcome by the use of immobilised aspartase (Tosa *et al,* 1973). Subsequently, the immobilised cell process has been shown to be 60% cheaper, chiefly because the cost of cell production has been reduced ten-fold (Fig. 6.1).

A single enzyme, aspartase, is involved. Fig. 6.5 gives the aspartase activities of various *Escherichia coli* preparations, and demonstrates that relatively little activity was lost on immobilisation of the cells. The immobilised cells had a much greater stability than either the immobilised enzyme or intact cells (Table 6.1), this stability decreasing as the operating temperature was raised. Thus, when used at 37,42 and 45°C the half-lives were 120,45 and 18 days respectively.

Escherichia coli cells were immobilised by dispersing them as a 20% (w/v) suspension in 15% (w/v) acrylamide at 20–25°C. The polyacrylamide gel was formed as a block which was then broken into 3–4 mm diameter granules. After immobilisation, the temperature optimum of the aspartase remained unchanged at 50°C, the pH optimum shifted from pH 10.5 to 8.5, and the enzyme lost its sensitivity to Mn^{2+} ions. Columns of the immobilised cell particles had zero order kinetics, and these kinetics were unaffected by changes in the column dimensions. On an industrial scale, 95% conversion was obtained by continuous use of the immobilised cells in sectionalised columns at 37°C with a space velocity of 0.6 (Sato *et al,* 1975). Mn^{2+} ions had to be added to the substrate to maintain the thermal stability of the column, the half-life being 120 days. Aspartic acid is collected by adjusting the eluate to pH2.8, crystallising, and then centrifuging or filtering the crystals before washing with water (Chibata *et al,* 1976a). Recently the Kyawa Hakko Co. have started a similar process (Yamada, 1978) and the Tanabe Co. are switching to the use of kappa-carageenan as their immobilisation support.

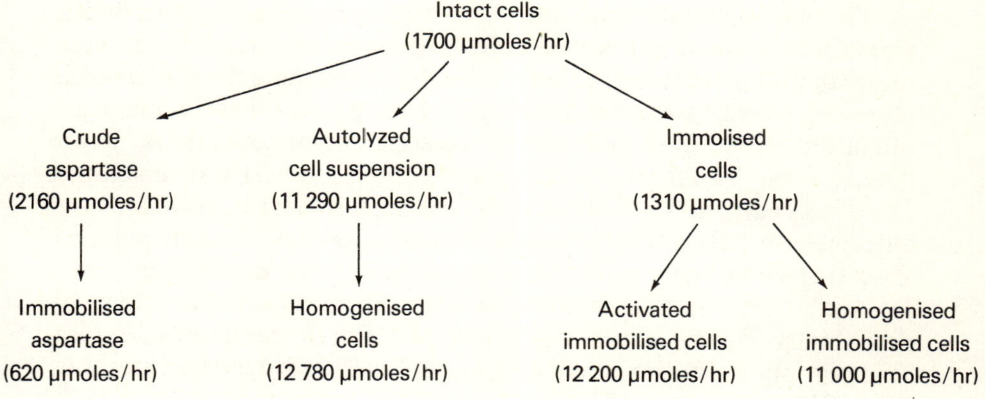

Fig. 6.5 – A comparison of the aspartase activities of various enzyme preparations per unit of initial cells. One gram (packed wet weight) of intact cells corresponds to 0.2g of dried cells. Numerical values in parenthesis are aspartase activities obtained from 1g of initial cells. (Chibata and Tosa, 1977).

L-malic acid which is used in the treatment of hepatic malfunctioning, especially hyperammonaemia, has been produced since 1974 by the Tanabe Seiyaku Co. using *Brevibacterium ammoniagenes* immobilised in polyacrylamide. The immobilised cells were used in continuous packed-bed columns using 1.0M sodium fumarate substrate at pH 7.0 and 37°C, pumped at 0.23 column volumes per hour. Approximately 15 tonnes of malic acid is produced per month in 70% yield. (Yamamoto *et al,* 1976). No changes in the pH or temperature optima of the enzyme occurred on immobilisation. Formation of the side product succinic acid was suppressed by treatment with a bile extract, which also increased the immobilised cell activity. Both effects are presumed to result from the bile acids destroying the chemiosmotic integrity of the cell membrane. When used in continuous columns at 37°C activity decays exponentially with a half-life of 52.5 days. The rate of loss of activity depends on the flow rate of substrate because of the leakage of enzyme and enzyme stabilisers from the immobilised cells, and because of poisoning by impurities in the substrate (Yamamoto *et al,* 1977).

As both the aspartic and malic acid processes require only single enzymes and have no co-factor and ATP requirements, then the main advantages of cell immobilisation do not as yet appear to have been industrially exploited. Neither process appears to use immobilised cells in the strictest sense, as in both cases the cells become ruptured after immobilisation (Chibata and Tosa, 1977); which raises the question of how the enzymes manage to be retained inside the lysed cells. Thus the first genuine industrial immobilised cell process appears to be the formation of prednisolone from Reichstein compound S via cortisol using the 11 β-hydroxylase and Δ^1 dehydrogenase activities of cells immobilised in polyacrylamide gel. This process has commenced operation in 1978 (Mosbach, K. personal communication). Some details of the 11 β-hydroxylation step were described by Mosbach and Larsson as early as 1970. Mycelia of the fungus *Curvularia lunata* were entrapped in polyacrylamide gel, no leakage of enzyme occurred, and after storage, activity could be regenerated by exposure to nutrients. For the Δ^1 dehydrogenation step, it has been calculated that 1.2 kg of polyacrylamide gel containing entrapped *Corynebacterium simplex* could supply the entire Swedish demand for dehydrogenated steroids. Thus 250 kg of steroid would be produced per year with a 1976 market price of £5M. In this process the Δ^1 dehydrogenase activity is greatly increased by the supply of 0.5% peptone dissolved in the substrate. This effect is due to cell division (7–10 fold) occurring within the polyacrylamide pellets, which causes a great increase in the concentration of immobilised cells, but induction of Δ' dehydrogenase activity is also likely (Fig. 6.6). The limited aqueous solubility of cortisol was overcome by use of the pseudocrystallofermentation technique, concentrations as high as 3,6g L^{-1} being achieved. In a later the increase in the Δ^1 dehydrogenase activity of *Arthobacter simplex* when induced with cortisol was proportional to the amount of microbial growth (Ohlson *et al,* 1979). This process is probably unrepresentative of cell immobilisation applications because of the very high value to bulk ratio of the product.

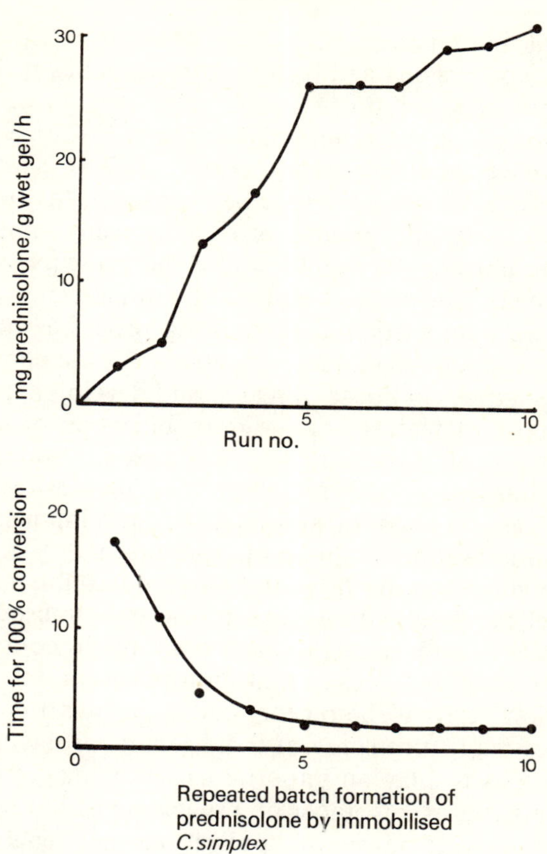

Fig. 6.6 – The repeated batch formation of prednisolone by immobilised *C. simplex*. 0.2g wet weight of *Corynebacterium simplex* gel was supended in 285 ml of 0.5% peptone pH 7.0 and 15 ml of 20mM cortisol in methanol. When 100% conversion to prednisolone was reached (indicated by the location of the points) the gel was again incubated with fresh cortisol containing medium. The whole experiment lasted 4 days (Larsson *et al*, 1976).

6.8 CONCLUSIONS

Immobilised cells have been used to catalyse the full range of enzymic reactions such as hydrolyses, oxidations, and decarboxylations, and the ability to carry out multi-enzyme conversions has been demonstrated. For any conversion the method of immobilisation and the mode of use will depend on economic, and a wide range of scientific and engineering factors, necessitating a multi-disciplinary approach. In particular the type of conversion, the yield and purity of the product and the activity and operational stability of the immobilised cells are very important parameters.

At present immobilised cells have been used chiefly to effect carbohydrate conversions and to form organic acids, including amino acids. As immobilised cell processes are often cheaper than the corresponding fermentation, immobilised enzyme or free enzyme processes (Yamamoto *et al,* 1974a; Yamamoto *et al,* 1974b; Sato *et al,* 1975; and Yamamoto *et al,* 1976) it may be possible in the future to carry out some previously uneconomic process by using immobilised cells and to invent novel processes to utilise the principal advantages of immobilised cells. The greatest chance of commercial success appears to be when no competition with existing processes occurs, that is when a novel product is manufactured, and when the commercial need has been satisfied by 'in house' research and development.

Many important features of the immobilisation of cells and their use need a great deal of further research. These include the alleviation of side reactions and further metabolism of the desired product, enhancment of gas and mass transfer rates, the development of efficient covalent immobilisation techniques, and the ability to utilise high molecular weight, water insoluble and water immiscible substrates. The leakage of cells and cell contents from the immobilised cell preparations must also be prevented, and it is desirable to control the *in situ* division of immobilised cells so that multiplication only occurs when the preparation needs to be regenerated. Continuation of the recent rapid progress may solve some of these general problems, and enable immobilised cells to play an important role in the future of the biochemical industries as a member of a spectrum of biological catalysts which covers a wide range of applications and enable a flexible approach to be taken to future problems.

ACKNOWLEDGEMENTS

I would like to thank the editors of Nature, Science, Advances in Applied Microbiology, Biotechnology and Bioengineering and the Journal of Fermentation Technology and the authors of Figs. 1–6 for their kind permission to reproduce their original material in this review.

REFERENCES

Abson, J. W. and Todhunter, K. H. (1967), in *Biochemical and Engineering Science* (ed. Blakeborough, N.) Academic Press, New York. Vol. 1, 309–343.

Arkles, B. and Briniger, W. (1975), *J. Biol. Chem.* **250**; 8856–8862.

Atkinson, B. (1974) in *Biochemical Reactors,* Pion Ltd, London, 191–214

Atkinson, B. and Daoud, I. S. (1976) in *Advances in Biochemical Engineering* (ed. Ghose, T. K., Fiechter, A. and Blakeborough, N.) Springer-Verlag. Berlin, Heidelberg and New York, Vol. 4, 90–92.

Bagnall, R. (1978), *Chem. in Ind.* **14**; 598-602.

Bang, W. G., Lang, S. and Sahm, H. and Wagner, F. (1978), *Proc 1st Eur. Cong. Biotechnol.* Part 1,2 186–189, Dechema Frankfurt F.R.G.

Betz, J. L., Brown, P. R., Smyth. M. J. and Clarke, P. H. (1974) *Nature* **247**; 261-264.

B. D. H. pamphlet *Cyanogum 41.*

Bucke, C. (1977) in *Topics in Enzyme and Fermentation Biotechnology,* ed. Wiseman, A., John Wiley, New York, London, Sydney and Toronto. Vol. 1 147–168.

Chang, T. M. S. (1971) *Nature* **229**; 117–118.

Cheetham, P. S. J., Blunt, K. W. and Bucke, C. (1979) *Biotechnol. Bioeng.,* submitted for publication.

Cheetham, P. S. J. (1979), *Enzyme and Microbial Technology,* **1**, 183–188.

Chibata, I., Tosa, T. and Sato, T. (1974a) *Appl. Micro.* **27**; 878–885.

Chibata, I., Kakimoto, T. and Nabe, K. (1974b) *Japanese Patent Kokai* **81**; 591/74.

Chibata, I., Kakimoto, T. and Nabe, K. (1974c) *Japanese Patent Kokai* **81**; 590/74.

Chibata, I., Tosa, T. and Sato, T. (1974d) *Japanese Patent Kokai* **132**; 290/74.

Chibata, I., Kato, J. Watanabe, T. and Uchida, T. (1975a) *Japanese Patent Kokai* **135**; 290/75.

Chibata, I., Tosa, T. Sato, T. and Yamamoto, K. (1975b) *Japanese Patent Kokai* **100**; 289/75.

Chibata, I., Tosa, T. and Sato, T. (1976a) in *Methods in Enzymology* (ed. Mosbach, K.), Academic Press, New York, San Francisco and London, Vol. 44, 739–746.

Chibata, I., Tosa, T., Sato, T. and Mori, T. (1976b) in *Methods in Enzymology* (ed. Mosbach, K.), Academic Press, New York, San Francisco and London, Vol. 44, 746–759.

Chibata, I., and Tosa, T. (1977) *Advances in Applied Microbiol* **22**; 1–25.

Chipley, J. B. (1974) *Microbiol.* **10**; 115–120.

Cohen, G. N. and Monod, J. (1957) *Bacteriol. Rev.* **21**; 169–174.

Cousinea, J. and Chang, T. M. S. (1977) *Biochem. Biophys. Res. Commun.* **79**; 24–31.

Dallyn, M., Falloon, W. C. and Bean, P. G. (1977) *Lab. Practise* **26**; 773–777.

Demain, A. L. (1971) *Adv. Biochem. Eng.* **1**; 113–118.

Dinelli, D. (1972) *Process Biochem.* **7**; 9–12.

Durrand, G. and Navarro, J. M. (1978) *Process Biochem.* **13**; 14–23.

Franks, N. E. (1971) *Biochem. Biophys. Acta.* **252**; 246–254.

Franks, N. E. (1978) in *Enzyme Engineering* (eds. Pye, E. K. and Weetall, H. H.), Plenum Press, London, New York. No. 3, 327–332.

Fujii, T., Matsumoto, K., Shibuya, Y., Hanamutsu, K., Yamaguchi, T., Watanabe, T. and Abe, S. (1974), British Patent 1,347,665.

Fukushima, S., Nagai, T. Fujita, K., Tanaka, A. and Fukui, S. (1978) *Biotechnol Bioeng* **20**; 1465–1469.

Ghose, T. K. and Subhash, C. (1978) *J. Ferm. Technol.* **56**; 315–322.

Goldstein, L. (1976) in *Methods in Enzymology* (ed. Mosbach, K.), Academic Press, New York, San Francisco, London, Vol. 44, 397–443.

Guttag, A. (1973) British Patent 1,426,101.

Hackel, U., Klein, J., Megnet, R. and Wagner (1975) *Eur. J. Appl. Micro.,* **1**; 291-293.

Hamilton, B. K., Montgomery, J. P. and Wang, D. I. C. (1974) in *Enzyme Engineering,* (eds. Pye, E. K. and Wingard, B.), Vol 2, 153–159. Plenum Press, New York and London.

Hattori, T. and Furusaka, C. (1961) *J. Biochem.* (Tokyo) **50**; 312–315.

Hattori, R. (1972) *J. Gen. Appl. Microbiol.* **18**; 319–325.

Haug, A., Larson, B. and Smidsröd, O. (1974) *Carbohydrate. Res.* **32**; 217–225.

Hirano, K., Karube, J., Otani, K. and Suzuki, S. (1975) *Ann. Meet. Soc. Ferm. Technol.* Japan., Osaka, p 249.

Hirano, K-I., Matsunaga, T. and Suzuki, S. (1977a) *J. Ferment. Technol.* **55**; 401–404.

Hirano, K., Karube, I., and Suzuki, S. (1977b) *Biotechnol. Bioeng.* **19**; 311–321.

Hough, J. S. and Lyons, T. P. (1972) *Nature* **235**; 389.

Horrisberger, M. (1971) *Biotechnol. Bioeng.* **13**; 589–584.

Horrisberger, M. (1976) *Biotechnol. Bioeng.* **18**; 1647–1651.

Ishimatzu, Y. Shigesada, S. and Kimura, S. (1976) Japanese Patent Kokai **86**; 142/76.

Jack, T. R. and Zajic, J. E. (1977) *Biotechnol. Bioeng.* **19**; 631–648.

Johnson, D. E. and Ciegler, A. (1969) *Biochem. and Biophys. Res. Commun.* **130**; 384–388.

Jones, J. B. (1976a) in Applications of Biochemical Systems in Organic Chemistry (Jones, J. B. , Sih, C. J. and Perlman, D., eds.), John Wiley, New York, Chapters 1 and 6.

Jones, J. B. (1976b) in *Methods in Enzymol* (ed. Mosbach, K.), Academic Press, NewYork, San Francisco and London Vol. 44, 831–844.

Kanno, T., Watanabe, H. and Sano, S. (1978) US patent 4, 106, 987.

Karube, I., Mitsuda, S., Matsunaga, T. and Suzuki, S. (1977a) *J. Ferment. Technol.* **55**; 243–248.

Karube, I., Matsunaga, T., Mitsuda, S. and Suzuki, S. (1977b) *Biotechnol. Bioeng.* **19**; 1535–1547.

Karube, I., Matsunaga, T., Tsura, S. and Suzuki, S. (1977c) *Biotechnol. Bioeng.* **19**; 1727–1733.

Karube, I., Hirano, K-I. and Suzuki, S. (1977d) *Biotechnol. Bioeng.* **19**; 1233–1238.

Kawabata, Y. and Demain, A. L. (1978) *Proc. of 176th ACS National Meeting (Microbiol and Biochemical Technology Division)* paper 48.

Kennedy, J. F., Barker, S. A. and Humphries, J. D. (1976) *Nature* **261**; 242–244.

Kierstan, M. and Bucke, C. (1977) *Biotechnol. Bioeng.* **19**; 387–397.

Kimura, A., Tatsutomi, Y., Mizushima, N., Tanaka, A., Matsuno, R. and Fukuda, H. (1978) *Eur. J. Applied. Microbial.* **5**; 13–16.

Kinoshita, S., Muranaka, M. and Okada, H. (1975) *J. Ferment. Technol.* **53**; 223–229.

Klein, J., Wagner, F., Washausen, P., Eng. H., and Martin, C. K. A. (1978) *Proc. 1st Eur. Cong. Biotechnol.* Part 1,2/190–193, Dechema Frankfurt, F. R. G.

Kobayashi, H. and Suzuki, H. (1976) *Biotechnol. Bioeng.* **18**; 37–51.

Kokubu, T., Karube, I. and Suzuki, S. (1978) *Eur. J. Appl. Microbiol* **5**; 233–240.

Kolarik, M. J., Chen, B. J., Emery, A. H. and Lin, H. C. (1974) in *Immobilised Enzymes in Food and Microbial Processes.* (ed. Olsen, A. C. and Cooney, C. L.), Plenum Press, New York, 71–83.

Kostner, A. and Mandel, M. (1976) in *Methods in Enzymology* (ed. Mosbach, K.), Academic Press, New York, San Francisco, London Vol. 44, 191–195.

Kumugaya, H. Sezima, S. Yamada, H. Hino, T. and Okamura, S. (1976), Annu. Meet. Agric. Chem. Soc. Jpn., Kyote, p233.

Kumakura, M. Yoshida, M. and Kaetsu, I. (1978) *Eur. J. Appl. Microbial.* **6**; 13–22.

Lagerlof, E., Nathorst-Westfelt, L., Ekstrom, B. and Sjoberg, B. (1976) in *Methods in Enzymology* (ed. Mosbach, K.) Academic Press, New York, San Francisco and London, Vol. 44, 759–768.

Larson, D. H. and Dimmick, R. (1964) *J. Bacteriol.* **88**; 1381–1413.

Larsson, P. O., Ohlson, S. and Mosbach, K. (1976) *Nature* **263**; 796–797.

Lawrence, R. L. and Okay, V. (1973) *Biotechnol. Bioeng.* **15**; 217–221.

Leuscher, F. German Patent 1, 227,855.

Linko, Y-Y., Pohjola, L. and Linko, P. (1977) *Process Biochem.* **12**; 14-16.

Long, M. E. (to Reynolds Tobacco Co.) (1977) US Patent 4,060,456.

Mansson, M. O., Mattiasson, B., Gestreluis, S. and Mosbach, K. (1976) *Biotechnol. Bioeng.* **18**; 1145–1159.

Marcipar, A. Cochet, N., Brackenbridge, L. and Lebeault, J. M. (1978) *Proc. 1st. Eur. Cong. Biotechnol.* Part 1,2/178–181, Dechema Frankfurt, F. R. G.

Marconi, W., Bartoli, F., Cecere, F., Galli, G. and Morisi, F. (1975) *Agr. Biol. Chem.* **39**; 277–279.

Martin, C. K. A. and Perlman, D. (1976a) *Biotechnol. Bioeng.* **18**; 217–237.

Martin, C. K. A. and Perlman, D. (1976b) *Eur. J. Appl. Micro.* **3**; 91–95.

Mavituna, F. and Sinclair, C. G. (1975) *Proc. 1st Eur. Cong. Biotechnol.* Part 1,2/182–185, Dechema, Frankfurt, F. R. G.

Meguro, S. and Konishi, H. (1977) *La Sucrerie Belge* **96**; 111–132.

Miles Laboratories Inc. (1974) British Patent 1,376,983.

Muyata, N., Kikuchi, T. and Furuichi, E. (1975) Annu. Meet. Agric. Chem. Soc. Jpn., Sapporo, p 58.

Mohan, R. R. and Li, N. N. (1974) *Biotechnol. Bioeng.* **16**; 513–523.

Mohan, R. R. and Li, N. N. (1975) *Biotechnol. Bioeng.* **17**; 1137–1156.

Mosbach, K. and Larson, P. O. (1970) *Biotechnol. Bioeng.* **12**; 19–27.

Murata, K., Tani, K., Kato, J., Chibata, I. (1978) *Eur. J. Appl. Microbiol.* **6**; 23–27.

Nakanishi, K., Adachi, S., Yamamoto, S., Matsuno, R., Tanaka, A. and Kamikubo, T. (1977) *Agric. Biol. Chem.* **41**; 2455–2462.

Navarro, J. M. and Durrand, G. (1977) *Eur. J. Applied. Micro.* **4**; 243–254.

Newman, H. N. (1974) *Microbios* **9**; 247–257.

Nishida, Y., Sato, T., Tosa, T. and Chibata, I. (1979) *Enz. and Microbial. Technol.* **1**; 95–99.

Obara, J., Hashimoto, S. (1976/7) *Sugar Technol. Review* **4**; 209.

O'Driscoll, K. F. (1976) in *Advances in Biochem. Eng.* (ed. Ghose, T. K., Fiecher, A. and Blakeborough, N.), Springer-Verlag, Berlin, Heidelberg, New York, Vol. 4, 156–172.

Ohlson, S., Larsson, P. O. and Mosbach, K. (1978) *Biotech. Bioeng.* **20**; 1267–1284.

Ohlson, S., Larsson, P. O. and Mosbach, K. *Eur. J. Applied Microbiol. and Biotechnol.* **7**; 103–110.

Ohmiya, K., Ohashi, H., Kabayashi, T. and Shimizu, S. (1977) *Applied and Envir. Microbiol.* **33**; 137–146.

Ollis, D. F. (1972) *Biotechnol. Bioeng.* **14**; 871–884.

Pache, W. (1978) *Eur. J. Applied Microbiol. and Biotech.* **5**; 171–176.

Packer, L. (1976) *Febs. Lett.* **64**; 17–19.

Petre, D., Noel, C., and Thomas, D. (1978) *Biotechnol. Bioeng.* **20**; 127–134.

Poulsen, P. and Zitton, L. (1976) in *Methods in Enzymology* (ed. Mosbach, K.), Academic Press, New York, San Francisco, London Vol. 44, 809–821.

Rechnitz, G. A., Riechel, T. L., Kobos, R K. and Meyerhoff, M. E. (1978) *Science* **199**; 440–441.

Regan, R. (1977) Stone & Webster Biochemical Symposium, Toronto, paper 2.

Saif, S., Tani, Y., and Ogata, K. (1975) *J. Ferm. Technol.* **53**; 380–386.

Sato, T., Mori, T., Tosa, T., Chibata, I., Kurui, M., Yamashita, K. and Sumi, A. (1975) *Biotechnol. Bioeng.* **17**; 1797–1804.

Sato, T., Tosa, T. and Chibata, I. (1976) *Eur. J. Applied Microbiol.* **2**; 153–160.

Savidge, T., Powell, L. W. and Warren, K. B. (1975) US Patent 3883394.

Schnarr, G. W. Szarek, W. A. and Jones, J. K. N. (1977) *Appl. and Envir. Micro.* **33**; 732–734.

Scott, C. D. and Hancher, C. W. (1976) *Biotechnol. Bioeng.* **18**; 1393–1403.

Senior, E., Bull, A. T. & Slater, J. H. (1976) *Nature* **263**; 476–479.

Shimzui, S., Morioko, H., Tani, Y. and Ogita, K. (1975) *J. Ferm. Technol.* **53**; 77–83.

Skryabin, G. K., Koshchenko, K. A., Surovtsev, U.I. and Mogilnitski G. M., Fikhte, B. A. and Tyurin, J. I. (1974) *J. Steroid Biochem.* **5**; 397.

Slowinski, W. and Charm, S. E. (1973) *Biotechnol. Bioeng.* **15**; 973–979.

Somerville, H. J., Mason, J. R. and Ruffell, R. N. (1977) *Eur. J. Appl. Micro.* **4**; 75–85.

Somerville, H. J. and Mason, J. R. (1978) *Biochem. Soc. Trans.* in press.

Srene, P. A., Mattiasson, B. and Mosbach, K. (1973) *Proc. Natl. Acad. Sci. USA* **70**; 1534–1538.

Takasaki, Y. (1969) in *Fermentation Advances* (ed. Perlman, D.) p 561.

Tanaka, A., Yasuhara, S., Gellf, G., Osumi, M. and Fukui, S. (1978) *Eur. J. Applied Biotechnol.* **5**; 17–27.

Takata, I., Tosa, T. and Chibata, I. (1977) *J. Solid Phase Biochem.* **2**; 225-236.

Thananunkel, D., Tanaka, M., Chichester, C. O. and Lee, T-C (1976) *J. Food Sci.* **41**; 173.

Thomas, T. D. and Batt, R. D. (1969) *J. Gen. Micro.* **58**; 371–380.

Toda, K. (1975) *Biotechnol. Bioeng.* **17**; 1729–1749.

Toda, K., and Shoda, M. (1975) *Biotechnol. Bioeng.* **17**; 481–497.

Tosa, T., Sato, T., Mori, T., Matu, Y. and Chibata, I. (1973) *Biotechnol. Bioeng.* **15**; 69–84.

Tosa, T., Sato, T., Mori, T. and Chibata, I. (1974) *Appl. Micro.* **27**; 886–889.

Tramper, J., Muller, F. and van der Plas, H. C. (1978) *Biotech. Bioeng.* **20**; 1507–1523.

Updike, S. M., Harris, D. R. and Shrago, E. (1969) *Nature* **224**; 1122–1123.

Venkatasubramanian, K., Vieth, N. R. and Constantinides, A. (1975) 75th Annu. Meet. Am. Soc. for Microbiol, Session 116.

Venkatasubramanian, K., Saini, R. and Vieth, W. R. (1974) *J. Ferm. Technol.* **52**; 268–278.

Venkatasubramanian, K. (1977) Sloan Webster Intl. Biochem. Symp., Toronto, paper 12.

Venkatasubramanian, K., Constantinides, A. and Vieth, W. R. (1978) in *Enzyme Engineering* Vol. 3, (Pye, E. K. and Weetall, H. H. eds.), Plenum Press, New York and London, pp 29–43.

Vieth, W. R., Wang, S. S., Saini, R. (1973) *Biotechnol. Bioeng.* **15**; 565–569.

Vieth, W. R. and Venkatasubramanian, K. (1976) in *Methods in Enzymology* (ed. Mosbach, K.) Academic Press, New York, San Francisco, London, Vol. 44, 768–776.

Voishvillo, N. E., Kamemitskii, A. V., Kahaikov, A. Ya., Leontev, I. G., Paukov, V. N. and Naknapetyan, L. A. (1976) I zv. Akad. Nauk. USSR Ser-Chem. 1303.

Wang, D. I. C. (1970) *Biotechnol. Bioeng.* **12**; 873–887.

van Wezel, A. L. (1967) *Nature* **216**; 64–65.

White, F. H. and Portno, A. D. (1978) *J. Inst. Brew.* **84**; 228–230.

Yamada, H. (1978) *Biotechnol. Bioeng.* **19**; 1563–1623.

Yamada, H., Yamada, M., Nagazawa, E , Kumagaya, H. Hino, T. and Okamura, S. (1975) Annu Meet. Agri. Chem. Soc. Jpn. Sapporo, p 336.

Yamamoto, K., Sato, T. Tosa, T. and Chibata, I. (1974a) *Biotechnol. Bioeng.* **16**; 1589–1599.

Yamamoto, K., Sato, T., Tosa, T. and Chibata, I. (1974b) *Biotechnol.* **16**; 1601–1610.

Yamamoto, K., Tosa, T., Yamashita, K. and Chibata, I. (1976) *Eur. J. Appl. Micro.* **3**; 169–183.

Yamamoto, K., Tosa, T., Yamashita, K. and Chibata, I. (1977) *Biotechnol. Bioeng.* **19**; 1101–1114.

Yang, H. S., Leung, K. H. and Archer, M. C. (1976) *Biotechnol. Bioeng.* **18**; 1425–1432.

Yang, H. S.,,and Studebaker, J. F. (1978) *Biotechnol. Bioeng.* **20**; 17–25.

Tasuda, H., Peterling, A., Colton, C. K., Smith, K. A. and Merrill, E. W. (1969) *Makromol. Chem.* **126**; 177–183.
Yasuda, H., Lamaze, C. E. and Peterling, A. (1971) *J. Polymer Sci.* **9**; 1117–1122.
Yoshino, M., Mashino, Y. and Morishita, M. (1975) US patent 3,875,000.

NOTE ADDED IN PROOF

Several recent, important developments in the field of immobilised cells must also be mentioned. In particular Brodelius *et al* (1979) have described the first successful immobilisation and use of plant cells. They used *Morinda citrifolia* cells to carry out the *de novo* synthesis of anthraquinones, *Catharanthus roseus* cells to form indole alkaloids, and *Digitalis laraba* cells to effect steriospecific hydroxylation of the cardiac glycoside digitoxin. The *Catharanthus* cells were able to recycle cofactors in situ, while the *Morinda* cells remained alive even after 22 days continuous use, and immobilisation appears to have diverted their metabolism away from growth and cell division and towards secondary product formation. The plant cells were entrapped in calcium alginate gels, a method which is distinguished by the mildness of the conditions required for immobilisation and so appears to be most appropriate for labile cells.

This same immobilisation technique has also been used to co-immobilise β-glucosidase and yeast cells. The enzyme-cell composite was used to metabolise an aqueous solution of cellobiose, the disaccharide derived from the hydrolysis of cellulose, to ethanol in almost theoretical yield. The hydrolysis of cellobiose to glucose was mediated by the β-glucosidase, and the metabolism of glucose to ethanol was carried out by the yeast cells; but the chief experimental difficulty involved optimising the conditions of use as the cell and enzyme have very different pH and temperature optima (IUPAC Congress in Helsinki).

At the same meeting Chibata described how his group have used yeast cells immobilised in carageenan and packed in columns, to obtain ethanol concentrations as high as 12.8% from a 20% (w/v) glucose solution supplemented with salts and nutrients. When a residence time of 2.5h was used the degree of conversion of glucose to ethanol was almost 100%, and the system was stable for at least 3 months continuous use.

Recently most advances have been made in providing a capability for metabolising water insoluble and immisicible substrates by using hydrophobic supports to immobilise cells. Two systems have been tested the Δ^1 dehydrogenation of hydrocortisone to prednisolone using *Arthobacter simplex*, and Δ^1 dehydrogenation of 4-androstene-3, 17-dione to androst-1,4–diene-3, 17-dione using *Nocardia rhodocrous*, both using substrate dissolved in benzene: heptane (1:1, *v/v*). In the former system the activities of the immobilised cells were 15 and 17% of the free cell activity using urethane and maleic polybutadiene as supports respectively. Using the Δ^1 dehydrogenase of *Nocardia*, immobilised cells 18% as active as free cells using urethane as support and only 23% as active using a support synthesised from hydroxyethylacrylate, isophorone diisocyanate and poly(propylene glycol)-2000 (ENTP-2000). Surprisingly the *Nocardia* cells immobilised in the ENTP-2000 were only slightly more active than when a hydrophilic support was used. This is because despite the apparent ease with which the steroid substrate and oxygen can penetrate into the support particles, diffusion of the hydrophilic artificial electron acceptor P.M.S., required for the dehydrogenation to take place, is very poor. The immobilised *Nocardia* cells were reported to have a greater operational stability than the free cells but were, of course, less active. (Omata *et al,* 1979; Sonomoto *et al,* 1979; Tanaka *et al,* 1979; and Yamané *et al,* 1979).

Despite the low effectiveness factors such heavily diffusion controlled systems must be of industrial importance as B.P. have patented the use of viable *Candida lipolytica* or *tropicalis*, or *Saccharomycopsis lipolytica* or *tropicalis* cells, entrapped in polyacrylamide or polyamide gels, to oxidise medium range normal paraffins ($C_{14}-C_{20}$). After entrapment the cells were packed in a column, a hydrocarbon-water emulsion recycled through the column and carboxylic acids assayed in the eluate (Lester, 1979).

Other immobilisation techniques involving novel synthetic support materials have also been reported from Japan. Hayashi *et al* (1979) have polymerised the sodium, potassium and magnesium salts of acrylic acid by exposure to radiation so as to immobilise *Brevibacterium ammoniagenes*. The cells NAD-kinase activity was then used to form NADP. Similarly Kumakura *et al* (1978) entrapped *Streptomyces phaechromogenes* cells by α-radiation induced polymerisation of 2-hydroxyethyl methacrylate at low temperatures. A very high concentration of cells could be immobilised on the surfaces of the polymer, no cell leakage was observed and substantial amounts of activity were retained.

Rather than use an immobilisation support, aggregation of cells by cross-linking or flocculation appears to be the immobilisation method of choice because of the high cell densities. However activity may be lost during cross-linking, internal diffusional restrictions inside the cell aggregate may be very high, and the aggregates are usually mechanically very weak so necessitating the use of strong, porous, support materials. Recently several attempts have been made to try to overcome these disadvantages, for instance Atkinson *et al* (1979) have devised a method of allowing cells to naturally adhere to and then grow on the surface of stainless steel wire knitted into open ball structures. These wire balls are packed into columns and used under low shear conditions with any excess cells being sloughed off the surface of the support; cell concentrations as high as 55g dry cells per litre pellet volume being obtained. The chief advantages claimed for this system are that even types of cells not normally regarded as being flocculant can be immobilised, mixed cell cultures can be established, and the method has excellent potential for scale-up. In common with other immobilisation methods the density of the cells is increased by immobilisation so that flow-through reactors can be operated at dilution rates much greater than the maximum specific growth rates of the micro-organisms.

A similar system appears to be the ceramic support developed by Messing and Oppermann (1978). At least 70% of the pores have diameters which range from the smallest dimension of the cell to less than five times the largest dimension of the cell. Also Corrieu *et al* (1978) have allowed films of yeast cells to grow on mixtures of Riesseguhr and p.v.c fragments. About 80 mg of active yeast was retained per g of support and it produced beer when wort was passed through the column.

REFERENCES

Atkinson, B., Black, G. M., Lewis, P. J. S. and Pinches, A. (1979) *Biotechnol. Bioeng.* **21**, 193–200.

Brodelius, P. Deus, B., Mosbach, K. and Zenk, M. H. (1979). *FEBS Lett.* **103**, 93–97.

Hayshi, T., Tanaka, Y. and Kawashina,, K. (1979) *Biotechnol. Bioeng.* **21**, 1019–1030.

Helsinki Meeting (IUPAC), Chem and Eng. News (1979), Sept. 17th, 27–29.

Kumakura, M., Yoshida, M. and Kaetsu, I. (1978) *Eur. J. Applied Microbiol.* **6**, 13–22.

Lester, D. E. (1979), British Patent No. 1 545 490 assigned to B.P. Co.

Messing, R. A. and Opperman, R. A. (1978), British Patent No. 2 004 300 assigned to the Corning Glass Works.

Omatia, T., Tanaka, A., Yamane, T. and Fukui, S. (1979) *Eur. J. Appl. Microbiol.* **6**, *207–215.*

Sonomoto, K., Tanaka, A., Omata, T., Yamane, T. and Fukui, S. (1979) *Eur. J. Appl. Microbiol.* **6**, 325–334.

Tanaka, A., Jin, I. N., Kawamoto, S. and Fukui, S. (1979) *Eur. J. Appl. Microbiol.* **7**, 351–354.

Yamané, T., Nakatani, H., Sada, E., Tanaka, A. and Fukui, S. (1979) *Biotechnol. Bioeng.* **21**, 2133–2145.

Index